The Foundations of Topological Graph Theory

Springer
New York
Berlin
Heidelberg
Barcelona
Budapest
Hong Kong
London
Milan
Paris
Santa Clara
Singapore
Tokyo

C. Paul Bonnington Charles H.C. Little

The Foundations of Topological Graph Theory

With 69 Illustrations

Springer

C. Paul Bonnington
Department of Mathematics
University of Auckland
Auckland, New Zealand

Charles H.C. Little
Department of Mathematics
Massey University
Palmerston North, New Zealand

Mathematics Subject Classification (1991): 05C10

Library of Congress Cataloging-in-Publication Data
Bonnington, C. Paul.
The foundations of topological graph theory/C. Paul Bonnington and Charles H.C. Little.
p. cm.
Includes bibliographical references and index.

ISBN-13: 978-1-4612-7573-2 e-ISBN-13: 978-1-4612-2540-9
DOI: 10.1007/978-1-4612-2540-9

1. Topological graph theory. I. Little, Charles H.C. (Charles Huw Crawford), 1947–. II. Title.
QA166.195.B66 1995
514'.223–dc20 95-30507

Printed on acid-free paper.

Softcover reprint of the hardcover 1st edition 1995

Production managed by Hal Henglein; manufacturing supervised by Jeffrey Taub.
Photocomposed copy prepared from the authors' LaTeX file.

9 8 7 6 5 4 3 2 1

Preface

This is not a traditional work on topological graph theory. No current graph or voltage graph adorns its pages. Its readers will not compute the genus (orientable or non-orientable) of a single non-planar graph. Their muscles will not flex under the strain of lifting walks from base graphs to derived graphs.

What is it, then? It is an attempt to place topological graph theory on a purely combinatorial yet rigorous footing. The vehicle chosen for this purpose is the concept of a 3-graph, which is a combinatorial generalisation of an imbedding. These properly edge-coloured cubic graphs are used to classify surfaces, to generalise the Jordan curve theorem, and to prove Mac Lane's characterisation of planar graphs. Thus they play a central role in this book, but it is not being suggested that they are necessarily the most effective tool in areas of topological graph theory not dealt with in this volume. Fruitful though 3-graphs have been for our investigations, other jewels must be examined with a different lens.

The sole requirement for understanding the logical development in this book is some elementary knowledge of vector spaces over the field $\mathcal{Z}_2$ of residue classes modulo 2. Groups are occasionally mentioned, but no expertise in group theory is required. The treatment will be appreciated best, however, by readers acquainted with topology. A modicum of topology is required in order to comprehend much of the motivation we supply for some of the concepts introduced.

We begin with a chapter on elementary graph theory. In keeping with the purpose of this book, we have tried to develop rigorously all of the ideas and results we need, no matter how intuitive they may seem. Many fundamental concepts – paths, connected graphs, isthmuses, circuits, trees, Eulerian graphs, bipartite graphs, n-connected graphs, cocycles – are characterised. In each case, we have chosen a definition which appeals to us because of its elegance and which, in our opinion, indicates the underlying reason for the importance of the concept being defined. The subsequent characterisation may then seem relatively mundane, but will usually be more familiar to many readers and is frequently more useful.

Chapter 2 introduces 3-graphs, and explains how they generalise imbeddings of graphs in surfaces. These purely combinatorial objects are not the only tools we could have chosen: other candidates have been supplied by Edmonds and by Tutte. We examine in considerable detail the precise relationship between their ideas and 3-graphs.

Surfaces are classified in Chapter 3, using 3-graphs. The key ideas emerging from this classification are the orientability and Euler characteristic of a 3-graph. These concepts are developed more fully in the next two chapters. The investigation of orientability in Chapter 4 leads us to consider coherently and consistently orientable families of circuits and bonds and takes us into the arcane world of forks and their tines. Some of the ideas re-emerge in Chapter 9. The study of the Euler characteristic in Chapter 5 yields a graph theoretic version of the theorem that the first Betti number of a surface is the largest number of closed curves that can be drawn in the surface without dividing it into two or more regions. This chapter also relates our work to the permutation pairs of Stahl.

The remainder of the book is largely devoted to planarity. Mac Lane's characterisation appears in Chapter 6 as a natural development of the ideas of the previous chapter and is used in the next to prove the famous Kuratowski characterisation. The book could reasonably have ended at this point, but the authors prove unable to curtail their verbosity and delve next into the topic of duality. This development leads inexorably to another characterisation of planarity, a generalisation of a theorem of Rosenstiehl and Read. Further characterisations of planarity occupy the final two chapters.

The patience of the authors' wives has been appreciated, and the authors also thank Springer-Verlag for their support. Any typographical errors are the result of the fumbling of the authors' fingers, and any mathematical errors arise from the bumbling of their brains.

Contents

1
Introduction

This chapter details the basic graph theoretical terminology and results that permeate the book.

1.1 Sets

Throughout this book, all sets are assumed finite unless an indication to the contrary is given. A set *contains* its elements and *includes* its subsets.

Let X and Y be sets. Their *union*, $X \cup Y$, and *intersection*, $X \cap Y$, are defined in the usual way. These operations are commutative and associative. We denote by $X - Y$ the collection of all elements of X not in Y. This set is the *complement* of Y in X. As in Graver and Watkins (1977), the *sum*, $X + Y$, of X and Y is their symmetric difference, $(X - Y) \cup (Y - X)$, but we do not eschew the use of the latter notation. This operation of addition is also commutative and associative. If X is a non-empty collection of sets, their union, intersection and sum are denoted by $\bigcup X$, $\bigcap X$ and $\sum X$, respectively. We also define $\bigcup \emptyset = \sum \emptyset = \emptyset$, where $\emptyset$ is the empty set. The cardinality of a set X is written as $|X|$. If X is a subset of Y, we write $X \subseteq Y$. We confine the use of the notation $X \subset Y$ to the case where the inclusion is proper. $\mathcal{P}(Y)$ denotes the collection of all subsets of Y. For any non-negative integer n, $\mathcal{P}_n(Y)$ denotes the collection of all $X \subseteq Y$ for which $|X| \leq n$.

A *pair* is a set of cardinality 2. A special case is the pair $\{x, \{x, y\}\}$ for any objects x and y: this is the *ordered pair* or *ordered 2-tuple* (x, y), also denoted by xy or $(yx)^{-1}$. Its *components* are x and y. If X and Y are sets, then the *Cartesian product* $X \times Y$ is the collection of all ordered pairs (x, y) for which $x \in X$ and $y \in Y$.

For any positive integer n, we write $J_n = \{1, 2, \ldots, n\}$ and $J_n^0 = J_n \cup \{0\}$. Given objects $x_1, x_2, \ldots, x_n$, where $n > 2$, we define the *ordered n-tuple* $(x_1, x_2, \ldots, x_n)$, also denoted by $x_1 x_2 \cdots x_n$ or $(x_n x_{n-1} \cdots x_1)^{-1}$, to be $(x_1 x_2 \cdots x_{n-1})x_n$. Its *components* are $x_1, x_2, \ldots, x_n$. An *ordered* 1-*tuple* is a single object, its only *component*. The empty set is the only *ordered* 0-*tuple* and has no components. An *ordered set* or *string* is an ordered n-tuple for some non-negative integer n. Its components also may be called *elements*. The notation $[x_1 x_2 \cdots x_n]$ means the set of all ordered n-tuples of the form $x_j x_{j+1} \cdots x_n x_1 x_2 \cdots x_{j-1}$, where $j \in J_n$. By abuse of language, we may say that $x_1, x_2, \ldots, x_n$ *belong* to $[x_1 x_2 \cdots x_n]$. If $A = x_1 x_2 \cdots x_k$ and $B = x_k x_{k+1} \cdots x_n$ for some $k \in J_n$, then we write $AB = x_1 x_2 \cdots x_n$. Thus

$$(AB)^{-1} = x_n x_{n-1} \cdots x_1 = (x_n x_{n-1} \cdots x_k)(x_k x_{k-1} \cdots x_1) = B^{-1} A^{-1}.$$

We also write $AB = B$ if $A = \emptyset$, and $AB = A$ if $B = \emptyset$.

We assume familiarity with the notion of a function and concomitant definitions. The domain of a function f is denoted $\mathcal{D}f$. The restriction of f to a subset X of $\mathcal{D}f$ is denoted by $f|_X$. If f and g are functions with disjoint domains, then $f \cup g$ is the function h, with domain $\mathcal{D}f \cup \mathcal{D}g$, such that $h|_{\mathcal{D}f} = f$ and $h|_{\mathcal{D}g} = g$. If $X \subseteq \mathcal{D}f$, then $f[X]$ denotes the set of all $f(x)$ such that $x \in X$. Given a set Y of ordered pairs, we may define $p(x, y) = x$ and $q(x, y) = y$ for each $(x, y) \in Y$. The ranges of the functions p and q so defined are the *first projection* and *second projection*, respectively, of Y.

A *partition*, $\mathcal{P}$, of a set X is a collection of non-empty disjoint subsets of X whose union is X. Thus, if $X = \emptyset$, then $\mathcal{P} = \emptyset$. The sets in $\mathcal{P}$ are the *cells* of the partition. A partition $\mathcal{Q}$ of X is a *refinement* of $\mathcal{P}$ if every cell of $\mathcal{Q}$ is a subset of a cell of $\mathcal{P}$. $\mathcal{P}$ is *trivial* if it has only one cell. If $Y \subseteq X$, then $\mathcal{P}|_Y$ is the partition of Y whose cells are the non-empty intersections with Y of the cells of $\mathcal{P}$.

A *pseudo-partition* of a set X is a collection of disjoint subsets of X whose union is X.

A *family* is an ordered pair of the form (X, f), where X is a set and f is a function mapping X into the set of all positive integers. For each $x \in X$ we call $f(x)$ the *multiplicity* of x. The elements of X are also called the *elements* or *components* of the family.

Let (X, f) and (Y, g) be families. We define $(X, f) \cup (Y, g)$ to be the family $(X \cup Y, h)$, where $h(x) = f(x)$ for each $x \in X - Y$, $h(y) = g(y)$ for each $y \in Y - X$, and $h(z) = f(z) + g(z)$ for each $z \in X \cap Y$.

Again, let (X, f) and (Y, g) be families. Let Z be the set of elements z of $X \cap Y$ for which $g(z) \geq f(z)$. Then we define $(X, f) - (Y, g)$ as the family $(X - Z, h)$, where $h(x) = f(x)$ for each $x \in X - Y$ and $h(y) = f(y) - g(y) > 0$ for each $y \in (X \cap Y) - Z$.

For any non-negative integer n, we may associate a family (X, f) with an ordered n-tuple $(x_1, x_2, \ldots, x_n)$ in the following way: We take $X = \bigcup_{i=1}^n \{x_i\}$, and for each $x \in X$ we define $f(x)$ to be the number of values of i in J_n such that $x_i = x$. The family (X, f) will often be written as $(x_1, x_2, \ldots, x_n)$, but it should be clear from the context whether the latter notation represents a family or an ordered n-tuple. The parentheses in this notation may occasionally be replaced by braces.

If $f(x) = 1$ for each $x \in X$, then the family (X, f) is frequently identified with the set X.

1.2 Graphs

We define a *graph* G as an ordered triple $(VG, EG, \psi G)$, where VG and EG are sets and ψG is a function mapping EG into $\mathcal{P}_2(VG) - \{\emptyset\}$. We call VG, EG, and ψG the *vertex set*, *edge set*, and *incidence function*, respectively, of G. The elements of VG and EG are the *vertices* and *edges*, respectively, of G. G is *null* if $VG = \emptyset$ and *empty* if $EG = \emptyset$.

We say that a graph *contains* its vertices and edges. Accordingly, we may describe the elements of $VG \cup EG$ as being *in* the graph G. The obvious extensions of this terminology to other graph theoretical concepts will be employed without being formally introduced.

Let G be a graph. An edge e is a *loop* or a *link* according to whether $|\psi G(e)| = 1$ or $|\psi G(e)| = 2$. The elements of $\psi G(e)$ are the *ends* of e, and e is *incident* on them and *joins* them. If v is the unique end of a loop e, then by abuse of language we also say that v is the *other* end of e. If v and w are the ends of e, we sometimes say that *e joins* $\{v, w\}$. (This convention enables us to say that a link joins a pair of vertices.) We may also say that *e joins v to w*, or vice versa. The ends of e in turn are *incident* on e. Two distinct edges are *adjacent* (to each other) if they are incident on a common vertex. Two distinct vertices are *adjacent* (to each other) if they are incident on a common edge. Adjacent edges and adjacent vertices are sometimes described as *neighbours* (of each other).

The *degree*, $\deg_G(v)$, in G of a vertex $v \in VG$ is $a + 2b$, where a and b, respectively, are the numbers of links and loops incident on v. We may delete the subscript in this notation if no ambiguity emerges. Vertices of degree 0 are *isolated*.

A graph may be represented in the plane by a drawing in which distinct vertices are represented by distinct points. If v' and w' are points that represent distinct vertices v and w, respectively, then an edge joining v and w is represented by a simple curve joining v' and w' and containing no other point that represents a vertex. If u' is a point representing a vertex u, then a loop incident on u is represented by a simple closed curve containing u' but no other point that represents a vertex. Thus, an object that is both a vertex and an edge is represented twice, once by a point and once by a curve.

Graphs G and H are *isomorphic* if there exist bijections $f : VG \to VH$ and $g : EG \to EH$ such that $\psi H(g(e)) = f[\psi G(e)]$ for all $e \in EG$. We say that (f, g) is an *isomorphism* from G to H, and that f and g are the *vertex* and *edge* isomorphisms, respectively. Clearly an equivalence relation is herein defined. Usually, one considers an equivalence class of isomorphic graphs rather than a single graph, since the internal structure of the vertices and edges escapes mention in the definition of a graph. It is therefore customary to identify the graphs in a particular equivalence class with a single representative selected from that class. Such a representative may be chosen to have its vertex set and edge set disjoint. Note that two isomorphic graphs may be represented by a common drawing. When a distinction between isomorphic graphs needs to be made, we denote by $[G]$ the class of all graphs isomorphic to a graph G. Thus, when we write $G = H$, we mean only that H is a graph in $[G]$.

1.3 Subgraphs

A graph H is a *subgraph* of a graph G if $VH \subseteq VG$, $EH \subseteq EG$, and $\psi H = \psi G|_{EH}$. We say that a graph *includes* its subgraphs and their edge sets. For any $T \subseteq EG$, $G - T$ denotes the subgraph $(VG, EG - T, \psi G|_{EG-T})$ and $G[T]$ the subgraph $(W, T, \psi G|_T)$, where W is the set of all vertices incident on an edge of T. The latter subgraph is *induced* by T. If $e \in EG$, then we sometimes say that G is obtained from $G - \{e\}$ by *adjoining* e. Now let $S \subseteq VG$. We define $G - S = (VG - S, U, \psi G|_U)$, where U is the collection of all edges not incident on a vertex in S. We also write $G[VG - S] = G - S$ and describe $G - S$ as the subgraph of G *induced* by $VG - S$. If v is an isolated vertex of G, then we say that G is obtained from $G - \{v\}$ by *adjoining* v. G is the *union*, $H \cup K$, of subgraphs H and K if $VH \cup VK = VG$ and $EH \cup EK = EG$.

Let H be a subgraph of G. H is a *proper* subgraph of G if $VH \cup EH \subset VG \cup EG$, and a *spanning* subgraph of G if $VH = VG$. A subgraph H of G that satisfies a specified property Z is a *minimal* subgraph of G satisfying Z if no proper subgraph of H satisfies Z. A subgraph H of G that satisfies a specified property Z is a *maximal* subgraph of G satisfying Z if H is not a proper subgraph of any other subgraph of G that satisfies Z.

1.4 Cocycles

We define the *cocycle* $\partial_G v$ of a vertex v in a graph G as the set of all links incident on v. The *cocycle* $\partial_G S$ of a set S of vertices is the sum of the cocycles of those vertices. Thus $\partial_G\{v\} = \partial_G v$; $\partial_G v$ is a *vertex cocycle*. The symbol ∂_G may be replaced by ∂ if no ambiguity results. Clearly, the edges of ∂S are those that join a vertex of S to a vertex of $VG - S$; hence $\partial S = \partial(VG - S)$. A set of edges is a *cocycle* if it is the cocycle of some set of vertices. An *isthmus* is the unique element of a cocycle of cardinality 1. A *bond* is a minimal non-empty cocycle. A graph G is *connected* if $\partial S \neq \emptyset$ for every non-empty proper subset S of VG and *bipartite* if EG is a cocycle. A *component* of G is a maximal connected subgraph. Thus the components of a non-null graph are non-null. Note that the word "component" has distinct meanings, but it should always be clear from the context which is intended. We write $c(G)$ for the number of components of G.

Let G be a graph and S, T disjoint subsets of VG. We denote by $[S, T]_G$ the set of edges joining a vertex of S to a vertex of T. Thus $[S, VG - S]_G = \partial_G S$. If v, w are distinct vertices of G, then we write $[\{v\}, \{w\}]_G$ as $[v, w]_G$. The subscript may be deleted throughout this notation if the reference to G is otherwise clear and no confusion is possible with other notation involving brackets.

1.5 Forests, trees, circuits, and paths

A *forest* is a graph in which every edge is an isthmus. A connected forest is a *tree*.

A *circuit* C in a graph G is the edge set of a minimal non-empty subgraph of G in which no edge is an isthmus. Its *length* is $|C|$. A circuit of length 1 has a loop as its unique element and will sometimes be called a *loop*. Circuits of length 2, 3, 4, 5, and n are called *digons*, *triangles*, *squares*, *pentagons*, and *n-gons*, respectively. Several authors would banish loops and even digons from the realm of graphs, but such a treatment is inappropriate for a book on topological graph theory. Indeed, loops and digons are important characters in our tale. Graphs that lack them are said to be *simple*. Given a graph G, the *skeleton* of G is the simple graph with vertex set VG such that distinct vertices are adjacent if and only if they are adjacent in G.

A *path* P joining vertices v and w in a graph G is the edge set of a minimal connected subgraph containing v and w. We may also say that P *joins* v to w, or vice versa. Its *length* is $|P|$. We call v and w the *ends* of P, and the edges of P incident on them are said to be *terminal edges* of P. The other vertices of $VG[P]$ are *internal vertices* of P, and the other edges of P are *internal edges* of P.

Let Q be a circuit or non-empty path in a graph G. We denote $VG[Q]$ by VQ. We say that Q *passes through* the elements of VQ, and we call VQ the *vertex set* of Q. Each element of Q is said to *carry* Q. We say that Q is *Hamiltonian* if $VQ = VG$.

A circuit C in a graph G is *induced* if $G[VC] = G[C]$.

1.6 Some elementary results

We now prove a number of elementary facts about the concepts we have introduced.

Lemma 1.1 *Let P be a path joining distinct vertices v and w in a graph G. Then the graph $G[P] - \{w\}$ is connected.*

Proof. Suppose not. Then there exists $S \subset VP - \{w\}$ such that $v \in S$ and $\partial_H S = \emptyset$, where $H = G[P] - \{w\}$. Subject to these conditions, choose S to minimise $|S|$. Let $K = G[P]$ and let $L = K[S \cup \{w\}]$. We shall prove that L is connected. Choose $T \subset VL$ such that $w \in T$, and suppose that $\partial_L T = \emptyset$. We distinguish two cases.

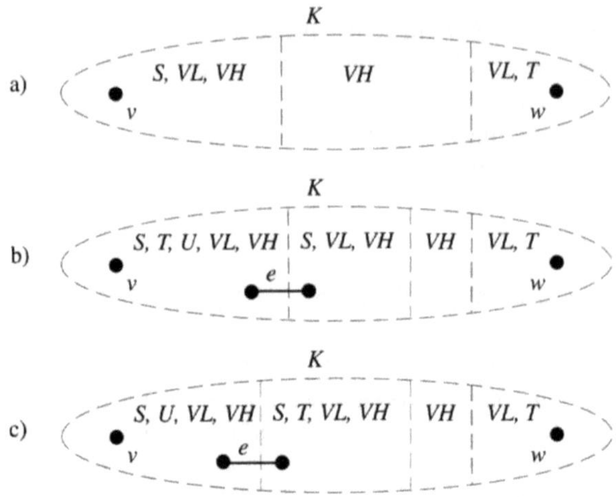

FIGURE 1.1.

Case 1. Suppose that $T = \{w\}$. (See Figure 1.1a.) Then $\partial_L S = \partial_L T = \emptyset$. But $\partial_H S = \emptyset$, and so $\partial_K S = \partial_H S \cup \partial_L S = \emptyset$. Moreover, $\emptyset \subset S \subset VP = VK$. Hence, we have the contradiction that K is not connected.

Case 2. Suppose that $T \neq \{w\}$. (See Figures 1.1b and c.) Let $U = S \cap T$ if $v \in T$ and let $U = S - T$ otherwise. Hence $v \in U$, and $U \subset S$ since $T \subset VL$ and $T \neq \{w\}$. By the minimality of $|S|$ it follows that $\partial_H U \neq \emptyset$. Choose $e \in \partial_H U$. Since $\partial_H S = \emptyset$, e must join a vertex of U to a vertex of $S - U$. Therefore, we have the contradiction that $e \in \partial_L T$ since $T \cap S \in \{U, S - U\}$.

We conclude that $\partial_L T \neq \emptyset$. Hence L is a connected graph containing v and w. Moreover, $EL \subseteq P$. By the minimality of P, it follows that $EL = P$. Hence $VP = VL = S \cup \{w\}$, in contradiction to the assumption that $S \subset VP - \{w\}$. ■

Theorem 1.2 *Let P be a path joining vertices v and w in a graph G. If $v \neq w$, then v and w are of degree 1 in $G[P]$ and every other vertex of VP is of degree 2 in $G[P]$. If $v = w$, then $P = \emptyset$.*

Proof. The theorem is clear if $v = w$, and so we suppose that $v \neq w$. If $|P| = 1$, the theorem is again immediate. Suppose therefore that $|P| > 1$ and that the theorem holds for all paths in G of smaller cardinality than P. Since $G[P]$ is connected, there exists $e \in \partial_{G[P]} w$. Let e join vertices w and x. By Lemma 1.1, $G[P] - \{w\}$ is a connected graph containing v and x. It therefore has a subgraph whose edge set Q is a path joining v and x. Certainly $Q \subset P$, since $e \notin Q$, and $v \neq x$ since $|P| > 1$. By assumption, therefore, $G[Q]$ is a connected graph in which v and x are of degree 1, and each vertex of $VQ - \{v, x\}$ is of degree 2 in $G[Q]$. Now define $R = Q \cup \{e\}$. Then v and w are of degree 1 in $G[R]$ and each vertex of $VG[R] - \{v, w\}$ is of degree 2 in $G[R]$.

We prove next that $G[R]$ is connected. Choose $S \subset VG[R]$ such that $w \in S$. If $S = \{w\}$, then $e \in \partial_{G[R]} S$. Suppose therefore that $S \neq \{w\}$. Let $T = S \cap VQ$.

Then $\emptyset \subset T \subset VQ$. Since $G[Q]$ is connected, it follows that $\partial_{G[Q]}T \neq \emptyset$. But $\partial_{G[Q]}T \subseteq \partial_{G[R]}S$, and so $\partial_{G[R]}S \neq \emptyset$. Hence $G[R]$ is connected.

We have $R \subseteq P$ by construction. Since $G[R]$ is a connected graph containing v and w, we infer from the minimality of P that $P = R$, and the theorem follows. ■

The following theorem provides a converse for Theorem 1.2.

Theorem 1.3 *Let v and w be distinct vertices of degree 1 in a connected graph G. Suppose that every vertex of $VG - \{v, w\}$ is of degree 2 in G. Then EG is a path joining v and w.*

Proof. Since G is connected, it has a subgraph whose edge set is a path P joining v and w. By Theorem 1.2, $\deg_{G[P]}(v) = \deg_{G[P]}(w) = 1$ and $\deg_{G[P]}(u) = 2$ for all $u \in VP - \{v, w\}$. From the hypotheses, it follows that $\partial_G VP = \emptyset$. Since G is connected, we deduce that $VG = VP$. Hence $G = G[P]$, and EG is the path P. ■

Let v and w be vertices or edges of a path P. Then we denote by $P[v, w]$ the edge set of the unique minimal connected subgraph of $G[P]$ containing v and w. $P[v, w]$ is a *subpath* of P.

Theorem 1.4 *A graph is connected if and only if every pair of vertices is joined by a path.*

Proof. Clearly, any pair of vertices of a connected graph must be joined by a path. Conversely, let G be a graph with that property. Let S be a non-empty proper subset of VG. Choose $v \in S$ and $w \in VG - S$, and let P be a path joining v and w. Let $T = VP \cap S$. Then $\emptyset \subset T \subset VP$. Since $G[P]$ is connected, it follows that $\partial_{G[P]}T \neq \emptyset$. But $\partial_{G[P]}T \subseteq \partial_G S$, and so $\partial_G S \neq \emptyset$. Hence G is connected. ■

From this theorem flow several corollaries.

Corollary 1.5 *Let K and L be connected subgraphs of a graph G. If $VK \cap VL \neq \emptyset$, then $G[EK \cup EL]$ is connected.*

Proof. If possible, choose $u \in VK - VL$, $v \in VK \cap VL$, and $w \in VL - VK$. $G[EK \cup EL]$ contains a path P joining u to v and a path Q joining v to w. Let x be the vertex of $VP \cap VQ$ that minimises $|VP[u, x]|$. Then $P[u, x] \cup Q[x, w]$ is a path in $G[EK \cup EL]$ joining u to w. Since any two vertices in K are joined by a path, and any two vertices in L are joined by a path, the result follows from Theorem 1.4. ■

Corollary 1.6 *Distinct components in a graph have disjoint vertex sets.*

Proof. Let K, L be distinct components in a graph G. If $VK \cap VL \neq \emptyset$, then $G[EK \cup EL]$ is connected. The maximality of K and L therefore yields the contradiction that each is equal to $G[EK \cup EL]$. ■

Corollary 1.7 *Let G be a graph and $\emptyset \subset S \subset VG$. Choose $v \in S$ and $w \in VG - S$. If $\partial S = \emptyset$, then v and w belong to distinct components of G.*

Proof. If not, then by Theorem 1.4 there is a path P in G joining v and w. Let $T = VP \cap S$. Then $\emptyset \subset T \subset VP$. Since $G[P]$ is connected, we have $\partial_{G[P]}T \neq \emptyset$. Hence, we obtain the contradiction that $\partial_G S \neq \emptyset$. ■

Corollary 1.8 *Let G be a graph and $S \subseteq VG$. Choose $e \in \partial S$, and let e join vertices v and w. Then v and w belong to distinct components of $G - \partial S$.*

Proof. Without loss of generality let $v \in S$ and $w \notin S$. Let $H = G - \partial S$. Then $\partial_H S = \emptyset$, and so v and w belong to distinct components of H by Corollary 1.7. ■

Corollary 1.9 *Let G be a connected graph. If $\emptyset \subset S \subset EG$, then there exists a vertex incident on an edge of S and on an edge of $EG - S$.*

Proof. Since $S \neq \emptyset$ and $EG - S \neq \emptyset$, we may choose a vertex v incident on an edge of S and a vertex w incident on an edge of $EG - S$. Suppose that neither vertex enjoys the desired property. They are joined by a path P since G is connected. Moreover, $P \cap S \neq \emptyset$ and $P \cap (EG - S) \neq \emptyset$, since $\partial v \subseteq S$ and $\partial w \subseteq EG - S$ by hypothesis. Therefore, some internal vertex of P satisfies the required property. ■

Corollary 1.10 *Let e be an isthmus in a connected graph G. Then $G - \{e\}$ has just two components.*

Proof. Since e is an isthmus, $G - \{e\}$ has more than one component, by Corollary 1.8. Let e join vertices v and w, and choose a component K of $G - \{e\}$ not containing v. Since G is connected, there is a path P in G joining v to a vertex $u \in VK$. Since $v \notin VK$, we must have $e \in P$. But then $P[u, w]$ is a path in $G - \{e\}$ joining u and w. Hence $w \in VK$. Thus, each component of $G - \{e\}$ contains v or w, and the result follows. ■

Conversely, if G is a connected graph and e is an edge that is not an isthmus, then $G - \{e\}$ is connected, for otherwise $\{e\}$ would be a cocycle of G.

The following theorem and its corollary are immediate from the definition of degree.

Theorem 1.11 *For any graph G,*

$$\sum_{v \in VG} \deg(v) = 2|EG|.$$

Corollary 1.12 *The number of vertices of odd degree in any graph is even.*

Theorem 1.13 *Let G be a graph, and let C be a non-empty subset of EG. Then C is a circuit if and only if $G[C]$ is connected and each vertex of $VG[C]$ is of degree 2 in $G[C]$.*

Proof. Suppose first that $G[C]$ is connected and each vertex of $VG[C]$ is of degree 2 in $G[C]$. Choose $e \in C$, and let e join vertices v and w.

Suppose that e is an isthmus of $G[C]$. By Corollary 1.8, v and w belong to distinct components, H and K, respectively, of $G[C]-\{e\}$. But $\deg_H(v)=1$ and $\deg_H(u)=2$ for each $u \in VH-\{v\}$. These results contradict Corollary 1.12. We conclude that no edge of C is an isthmus of $G[C]$.

Next, choose a non-empty proper subset X of C. Suppose that every vertex of $VG[X]$ is of degree 2 in $G[X]$. It follows from the hypotheses that $\partial_{G[C]}VG[X]=\emptyset$. Since $G[C]$ is connected, we infer that $VG[X] = VG[C]$. The hypothesis concerning the degrees of the vertices of $VG[X]$ then yields the contradiction that $X=C$. Since the vertices of $VG[X]$ are of positive degree in $G[X]$, one of them must therefore be of degree 1 in $G[X]$. The edge of X incident on this vertex is an isthmus of $G[X]$. We conclude that every non-empty proper subgraph of $G[C]$ must have an isthmus. Therefore C is a circuit.

Conversely, suppose that C is a circuit. The minimality of $G[C]$ implies that $G[C]$ is connected. Choose $e \in C$. Then $G[C]-\{e\}$ is connected, since e is not an isthmus. Therefore, there is a path P in $G[C]-\{e\}$ joining the ends of e. Let $R=P\cup\{e\}$. It follows by Theorem 1.2 that every vertex of $VG[R]$ is of degree 2 in $G[R]$. Moreover, $G[R]$ is connected by Corollary 1.5. It follows that R is a circuit. But $R\subseteq C$. Hence, $R=C$ by the minimality of C, and the theorem follows. ■

Let $\{e_1, e_2, \ldots, e_n\}$ be a circuit C, where e_i is adjacent to e_{i+1} for each $i \in J_{n-1}$. (It follows that e_n is adjacent to e_1.) Then the sets $[e_1e_2\cdots e_n]$ and $[e_ne_{n-1}\cdots e_1]$ are *orbits* of C.

Theorem 1.14 *An edge is an isthmus if and only if it belongs to no circuit.*

Proof. Let C be a circuit in a graph G, and choose $e \in C$. If e is a loop, it cannot be an isthmus, and so we assume that e is a link. Choose $S \subset VG$ so that $e \in \partial_G S$. Let $T = S\cap VC$. Then $e \in \partial_{G[C]}T$, and so $|\partial_{G[C]}T| > 1$ since e is not an isthmus of $G[C]$. Since $\partial_{G[C]}T \subseteq \partial_G S$, it follows that $|\partial_G S| > 1$ and that e is not an isthmus of G.

Conversely, suppose that an edge e is not an isthmus in a graph G. Let K be the component of G containing e. Then $K-\{e\}$ must be connected. Therefore, there is a path P in $K-\{e\}$ joining the ends of e. The set $P\cup\{e\}$ is then a circuit of G containing e. ■

Let v be a vertex of a circuit C, of length greater than 1, in a graph G. Let a, b be the edges of C incident on v; neither edge is an isthmus of $G[C]$. Therefore $G[C-\{a\}]$ is connected, and so $C-\{a\}$ is a path by Theorem 1.3. By Lemma 1.1 and Theorem 1.3 it follows that $C-\{a,b\}$ is also a path. We denote this path by C_v.

1.7 Theorems about trees

Theorem 1.15 *A connected graph is a tree if and only if it has no circuit.*

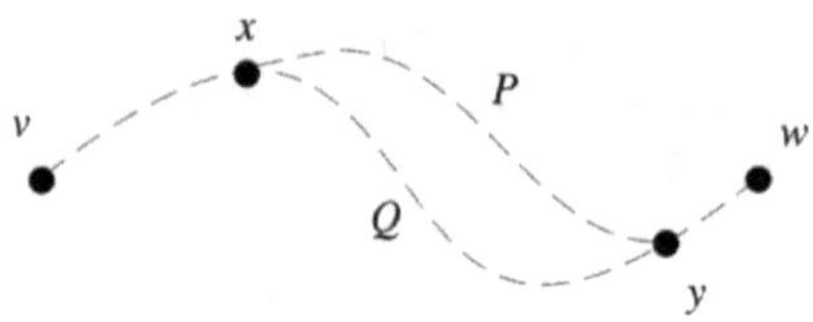

FIGURE 1.2.

Proof. Since every edge of a tree is an isthmus, the result is immediate from Theorem 1.14. ■

Theorem 1.16 *A connected graph with no loops is a tree if and only if every pair of vertices is joined by a unique path.*

Proof. Suppose that G is a connected graph with no loops but is not a tree. Then, by Theorem 1.15, G has a circuit C with $|C| > 1$. Any pair of vertices in VC is joined in G by more than one path.

On the other hand, suppose that G is a tree. Since G is connected, every pair of vertices is joined by a path. Suppose that there exist distinct vertices v and w joined by distinct paths P and Q. (See Figure 1.2.) Then there exist distinct vertices $x, y \in VP \cap VQ$ such that $P \cap Q[x, y] = \emptyset$. We thus have the contradiction that $P[x, y] \cup Q[x, y]$ is a circuit in G. Therefore, v and w are joined in G by a unique path. ■

Lemma 1.17 *If G is a connected graph, then $|EG| \geq |VG| - 1$.*

Proof. The lemma holds if $|VG| \leq 1$. Suppose that $|VG| > 1$ and that the lemma holds for each graph with fewer than $|VG|$ vertices. Choose $v \in VG$, and let $H = G - \{v\}$. Let $H_1, \ldots, H_k$ be the components of H. For each i we have $\partial_G VH_i \neq \emptyset$, but $\partial_H VH_i = \emptyset$, and so there is an edge of $\partial_G VH_i$ incident on v. Hence, the induction hypothesis gives

$$\begin{aligned} |EG| &\geq \sum_{i=1}^{k} |EH_i| + |\partial_G v| \\ &\geq \sum_{i=1}^{k} (|VH_i| - 1) + k \\ &= \sum_{i=1}^{k} |VH_i| \\ &= |VG| - 1. \blacksquare \end{aligned}$$

Theorem 1.18 *A non-null connected graph G is a tree if and only if*

$$|EG| = |VG| - 1.$$

Proof. First, let G be a tree. The theorem holds if $EG = \emptyset$. Suppose therefore that $|EG| > 0$ and that the theorem holds for all trees with fewer than $|EG|$ edges. Choose $e \in EG$. Since e is an isthmus, Corollary 1.10 shows that $G - \{e\}$ has just two components, K and L, both of which are trees. By the induction hypothesis it follows that

$$\begin{aligned} |EG| &= |EK| + |EL| + 1 \\ &= (|VK| - 1) + (|VL| - 1) + 1 \\ &= |VG| - 1. \end{aligned}$$

Conversely, suppose that $|EG| = |VG| - 1$ and that G has a circuit C. Choose $e \in C$. Then $G - \{e\}$ is connected by Theorem 1.14. But

$$\begin{aligned} |E(G - \{e\})| &= |EG| - 1 \\ &= |VG| - 2, \end{aligned}$$

in contradiction to Lemma 1.17. Hence G is a tree. ■

Corollary 1.19 *A tree with more than one vertex has at least two vertices of degree* 1.

Proof. If G is a tree with $|VG| > 1$, then the degree of each vertex is at least 1 since G is connected. But $\sum_{v \in VG} \deg(v) = 2|EG| = 2|VG| - 2$, and so at least two vertices are of degree 1. ■

1.8 Spanning trees

A *spanning tree* in a graph G is a spanning subgraph of G that is a tree. Only connected graphs have them.

Theorem 1.20 *A graph has a spanning tree if and only if it is connected.*

Proof. It follows from Theorem 1.4 that a graph that is not connected cannot have a connected spanning subgraph. Thus it has no spanning tree.

Suppose now that G is a connected graph. The theorem holds if $EG = \emptyset$, and so we assume as an inductive hypothesis that every connected graph with fewer than $|EG|$ edges has a spanning tree. If G itself is not a tree, it has a circuit C by Theorem 1.15. Choose $e \in C$. Since e is not an isthmus by Theorem 1.14, it follows that $G - \{e\}$ is connected and therefore has a spanning tree T. T is then a spanning tree of G also. ■

A *spanning forest* of a graph G is a spanning subgraph that is a forest such that each component is a spanning tree of a component of G. Thus, it is not merely a spanning subgraph that is a forest.

Let T be a spanning tree of a connected graph G, and let $e \in EG - ET$. By Theorem 1.16, ET includes a unique path P joining the ends of e. Thus $P \cup \{e\}$

is a circuit. We call this circuit a *fundamental* circuit of G (with respect to T or ET) and denote it by $T(e)$.

1.9 Eulerian graphs

A graph G is *Eulerian* if there is a partition of EG whose cells are circuits.

Theorem 1.21 *A graph is Eulerian if and only if the degree of every vertex is even.*

Proof. The theorem clearly holds for Eulerian graphs. Conversely, let G be a graph in which each vertex has even degree. If $EG = \emptyset$, the theorem holds vacuously. Suppose now that $EG \neq \emptyset$ and that the theorem holds for graphs with fewer edges than G. Since G has no vertices of degree 1, by Corollary 1.19 it cannot be a forest and so contains a circuit C. Every vertex of the graph $G - C$ is of even degree, and hence there is a partition $\mathcal{P}$ of $EG - C$ whose cells are circuits. The required partition of EG is $\mathcal{P} \cup \{C\}$. ■

1.10 Bipartite graphs

Let v, w be vertices in the same component of a graph G. By Theorem 1.4, there exists a path joining v and w. The length of a shortest such path is the *distance* from v to w in G. If $v, w \in VD$, where D is a path or circuit in G, then the *distance* from v to w in D is the distance from v to w in $G[D]$.

The proof of the next theorem uses the following facts: If $S_1, S_2, \ldots, S_n$ and T are sets, then

$$|\sum_{i=1}^{n} S_i| \equiv \sum_{i=1}^{n} |S_i| (\text{mod } 2)$$

and

$$T \cap \sum_{i=1}^{n} S_i = \sum_{i=1}^{n} (T \cap S_i).$$

Theorem 1.22 *A graph is bipartite if and only if the length of every circuit is even.*

Proof. In this proof, all congruences are taken modulo 2.

The edge set of a bipartite graph is a cocycle, and so the length of every circuit is even. Conversely, suppose that every circuit of a graph G has even length. It suffices to consider the case where G is connected.

Choose $v \in VG$. Let S be the set of vertices at an even distance from v. Suppose that vertices x and y of S are joined by an edge e. Let P and Q be paths of even length joining v to x and y, respectively. Then

$$\begin{aligned} |P + Q + \{e\}| &\equiv |P| + |Q| + 1 \\ &\equiv 1. \end{aligned}$$

But for any $w \in VG$ we have

$$\begin{aligned} \deg_{G[P+Q+\{e\}]}(w) &\equiv |(P+Q+\{e\}) \cap \partial w| \\ &\equiv |P \cap \partial w| + |Q \cap \partial w| + |\{e\} \cap \partial w| \\ &\equiv 0. \end{aligned}$$

By Theorem 1.21 it follows that $G[P + Q + \{e\}]$ is a Eulerian graph with an odd number of edges. We therefore obtain the contradiction that this graph has a circuit of odd length. Hence, no two vertices of S are adjacent, and similarly, no two vertices of $VG - S$ are adjacent. It follows that $EG = \partial S$, and G is bipartite. ■

If G is a bipartite graph and S is a subset of VG such that $EG = \partial S$, then we say that $\{S, VG - S\}$ is a *bipartition* of G. Note that vertices in S are not adjacent, and neither are vertices in $VG - S$. In other words, if the vertices in S and those in $VG - S$ are painted black and white, respectively, then every edge of G joins a black vertex to a white one.

1.11 Contractions

Let G be a graph and $\mathcal{Q}$ a partition of VG such that $G[Q]$ is connected for each $Q \in \mathcal{Q}$. Let H be a graph with vertex set $\mathcal{Q}$, and let EH be the set of edges of G that join vertices in distinct members of $\mathcal{Q}$. Suppose that each edge of H joins the two elements of $\mathcal{Q}$ containing its ends in G. Then H is called the *vertex contraction* (or simply *contraction*) of G determined by $\mathcal{Q}$. (See Figure 1.3.) Let $S \subseteq EG$ and $T \subseteq EH$. We say that S *contracts* to T in H, or that T is a *contraction* of S in H, if H is a vertex contraction of G and $T = S \cap EH$. We denote T by S_H. If $e \in EG$, $a \in EH$, and $\{e\}$ contracts to $\{a\}$ in H, then we say that e *contracts* to a or that a is a *contraction* of e. In this case, we write $a = e_H$.

Let e be a link in G joining vertices v and w. Let $\mathcal{Q}$ be the partition of VG such that $\{v, w\} \in \mathcal{Q}$ and $|\mathcal{Q}| = |VG| - 1$. Then the vertex contraction of G determined by $\mathcal{Q}$ is denoted by G_e. The vertices of G_e are therefore subsets of VG, by definition. A unique vertex of G_e is of cardinality 2; the cardinality of the others is 1.

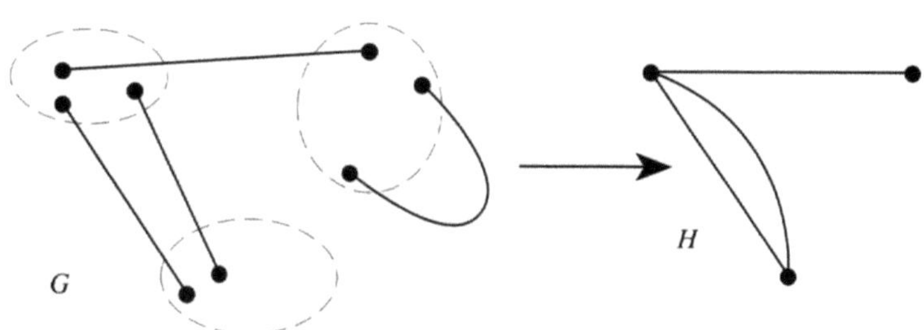

FIGURE 1.3.

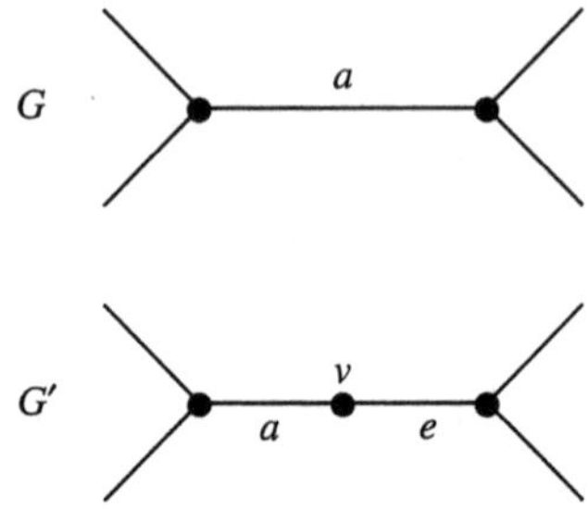

FIGURE 1.4.

Let G and G' be graphs. We say that G' is an *elementary subdivision* of G if there is a link $e \in EG'$, incident on a vertex v of degree 2, such that G is isomorphic to the contraction G'_e. (See Figure 1.4.) If a is the edge of $G' - \{e\}$ incident on v, then $a \in EG$. We say that G' is obtained from G by *subdividing a* with v. A graph H is said to be a *subdivision* of G if for some non-negative integer n there exist graphs $G_0, G_1, \ldots, G_n$ such that $G_0 = G$, $G_n = H$, and G_i is an elementary subdivision of G_{i-1} for each $i \in J_n$.

The concept of a contraction is helpful in obtaining the following characterisation of cocycles.

Theorem 1.23 *Let G be a graph and let $S \subseteq EG$. Then S is a cocycle if and only if $|S \cap C|$ is even for every circuit C.*

Proof. If S is a cocycle, it is immediate that $|S \cap C|$ is even for every circuit C.

Conversely, suppose that $|S \cap C|$ is even for every circuit C. Let K be a component of $G - S$, and suppose that there exists $e \in EG[VK] \cap S$. Let e join vertices v and w. By Theorem 1.4, there is a path P in K joining v and w. Since $e \in S$, we have $e \notin EK$, and so $P \cup \{e\}$ is a circuit. This conclusion is a contradiction, because $|S \cap (P \cup \{e\})| = 1$.

Now let $L_1, L_2, \ldots, L_m$ be the components of $G - S$. Let H be the vertex contraction of G determined by $\{VL_1, VL_2, \ldots, VL_m\}$. Since $EG[VL_i] \cap S = \emptyset$ for each $i \in J_m$, we find that $|S_H \cap C_H| = |S \cap C|$ for each circuit C in G. Moreover, $S_H = EH$ and so $|C_H| = |S \cap C|$, which is even. Hence, by Theorem 1.22, H is bipartite since any circuit of H is a contraction of a circuit of G. Therefore $S_H = \partial_H T$ for some $T \subseteq VH$. It follows that $S = \partial_G(\bigcup T)$. ■

1.12 Menger's theorem and n-connected graphs

Let G be a graph with distinct non-adjacent vertices v and w. A *vw-separating* set is a set $S \subseteq VG - \{v, w\}$ such that v and w belong to distinct components of $G - S$. Thus any subset of $VG - \{v, w\}$ is vw-separating if v and w are in distinct components of G. Paths P and Q joining v and w are said to be *independent* if

$VP \cap VQ = \{v, w\}$. A set of paths joining v and w is *independent* if any two paths in the set are independent.

The ideas above are related via a famous theorem of Menger. Proofs of this theorem abound; the one presented below is due to McCuaig (1984).

Theorem 1.24 (Menger) *Let v and w be distinct non-adjacent vertices in a graph G. Then the maximum cardinality M of an independent set of paths joining v and w is equal to the minimum cardinality m of a vw-separating set.*

Proof. Clearly $M \le m$. It remains only to establish the existence of m independent paths joining v and w when there is no vw-separating set of cardinality less than m. The proof is by contradiction.

Suppose that $m = 1$. Then v and w belong to the same component of G, and the required path exists by Theorem 1.4.

Now suppose that $m > 1$. Since there is no vw-separating set of cardinality less than $m - 1$, we may assume as an inductive hypothesis that there is an independent set $\mathcal{P}$ of $m - 1$ paths joining v and w. Let $H = G[\bigcup \mathcal{P}]$. Note that the set A of $m - 1$ vertices of VH adjacent to v cannot be a vw-separating set in G. Thus v and w belong to the same component of $G - A$, and so, by Theorem 1.4, there must be a path P in G joining v to a vertex of $VH - \{v\}$ whose edge incident on v is not in EH. Choose $x \in (VP - \{v\}) \cap VH$ so as to minimise $|P[v, x]|$. (Such an x exists because P joins v to a vertex of $VP \cap VH$.) We may assume that the pair $(\mathcal{P}, P)$ is chosen to minimise the distance in $G - \{v\}$ between x and w. We may also assume that $x \ne w$, for otherwise $\mathcal{P} \cup \{P\}$ is the required set of paths.

In $G - \{x\}$ there is no vw-separating set of cardinality less than $m - 1$. By induction there must therefore be an independent set $\mathcal{Q}$ of $m - 1$ paths in $G - \{x\}$ joining v and w. We choose $\mathcal{Q}$ to minimise $|\bigcup \mathcal{Q} \cap B|$, where $B = EG - (EH \cup P[v, x])$. Let $K = G[\bigcup \mathcal{Q}]$. Hence $x \notin VK$. Since $|\mathcal{Q}| < m$, there exists a path $Y \in \mathcal{P} \cup \{P[v, x]\}$ whose edge incident on v is not in EK. Choose $y \in (VY - \{v\}) \cap (VK \cup \{x\})$ to minimise $|Y[v, y]|$. This vertex exists because $w \in VY \cap VK$ unless $x \in VY$. We shall prove that $y = w$.

Suppose that $y = x$. (See Figure 1.5.) Let R be a shortest path in $G - \{v\}$ joining x and w. Choose $z \in VR \cap VK$ so as to minimise $|R[x, z]|$. Thus $z \ne x$, and so the distance in $G - \{v\}$ between z and w is less than $|R|$. For any path $Q \subseteq Y[v, x] \cup R[x, z]$ joining v and z, we have $(VQ - \{v\}) \cap VK = \{z\}$, and so the choice of $(\mathcal{P}, P)$ is contradicted by the pair $(\mathcal{Q}, Q)$. Hence $y \ne x$.

Next, suppose that $y \in VK - \{w\}$. Then there exists $X \in \mathcal{Q}$ such that $y \in VX$. Let u be the vertex of $VX \cap (VY - \{v\})$ that minimises $|X[v, u]|$. As the edge of X incident on v is not in Y, neither is the edge of $X[v, u]$ incident on u. Since $Y \in \mathcal{P} \cup \{P[v, x]\}$, the latter edge must therefore be in B, for otherwise the vertex sets of two paths in $\mathcal{P} \cup \{P[v, x]\}$ would share the vertex $u \notin \{x, v, w\}$. Hence $B \cap X[v, y] \ne \emptyset$. Now let $X' = Y[v, y] \cup X[y, w]$, and define $\mathcal{Q}' = (\mathcal{Q} - \{X\}) \cup \{X'\}$. Then $\mathcal{Q}'$ is an independent set of $m - 1$ paths in $G - \{x\}$ joining v and w, yet $|\bigcup \mathcal{Q}' \cap B| < |\bigcup \mathcal{Q} \cap B|$. This contradiction to the choice of $\mathcal{Q}$ shows that $y = w$.

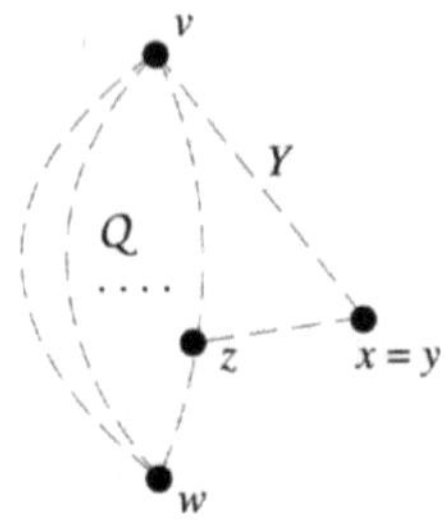

FIGURE 1.5.

We conclude that $\mathcal{Q} \cup \{Y\}$ is the required set of paths. ∎

Let n be a positive integer. A graph is said to be *n-connected* if any two distinct vertices are joined by at least n independent paths. Thus, a graph is 1-connected if and only if it is connected. Note that an n-connected graph G is m-connected for all $m \in J_n$. Moreover, if a spanning subgraph of G is n-connected, then so is G. Observe also that any graph G is n-connected if and only if $G - S$ is n-connected, where S is the set of loops of G.

Menger's theorem yields the following characterisation of simple n-connected graphs with more than one vertex.

Theorem 1.25 *A simple graph G with more than one vertex is n-connected if and only if $n < |VG|$ and $G - S$ is connected for each $S \subseteq VG$ such that $|S| < n$.*

Proof. Suppose first that G is n-connected. Any two distinct vertices of G are joined by at least n independent paths. Since G is simple, at most one of these paths can be of length 1, and so we must have $|VG| > n$. Choose $S \subseteq VG$ such that $|S| < n$, and suppose that $G - S$ is not connected. Let K, L be components of $G - S$, and choose vertices $u \in VK$ and $v \in VL$. Any path in G joining u and v must pass through a vertex of S, and so we have the contradiction that there cannot be n independent paths joining u and v. Therefore $G - S$ is connected.

Conversely, suppose that $n < |VG|$ and that $G - S$ is connected for each $S \subseteq VG$ such that $|S| < n$. Suppose that x and y are distinct vertices in G joined by fewer than n independent paths. Then x and y are joined by an edge e, for otherwise Menger's theorem reveals the existence of a subset S of VG such that $|S| < n$ and x and y lie in distinct components of $G - S$. In the graph $H = G - \{e\}$, fewer than $n - 1$ independent paths join x to y, and x and y are not adjacent. Hence there is a subset U of VG such that $|U| \leq n - 2$ and x and y lie in distinct components, X and Y respectively, of $H - U$. These are the only components of $H - U$, for otherwise $G - U$ could not be connected. Since $|VH| \geq n + 1$, it follows that $|VH - U| \geq 3$, and so we may assume without loss of generality that $|VX| \geq 2$. It follows that the graph $G - (U \cup \{x\})$ is not connected, contrary to hypothesis. We conclude that G is n-connected. ∎

1.13 2-connected graphs

We have already seen that the 1-connected graphs are the connected ones. The following theorem gives a useful characterisation of 2-connected graphs.

Theorem 1.26 *A connected graph G, is 2-connected if and only if any two links belong to a common circuit.*

Proof. Suppose that any two links of G belong to a common circuit. Let u, v be distinct vertices of G. Since G is connected, we may select links e and a incident on u and v, respectively. By hypothesis, e and a belong to a common circuit C, which therefore passes through u and v. Hence u and v are joined by two independent paths whose union is C.

Conversely, suppose that G is 2-connected. Let e, a be distinct links of G. Each of these links belongs to a circuit, since the ends of each are joined by at least two independent paths. It remains to show that there exists a circuit containing both e and a. Let e join vertices w and x, and let a join vertices y and z. Let H be the graph obtained from G by subdividing e and a with u and v, respectively.

We prove that H is 2-connected. For this purpose, it suffices to inspect its skeleton H', which is a simple spanning subgraph. Note that H' is connected since G is connected. Moreover, $G - \{t\}$ is connected for each $t \in VG$, for if $G - \{t\}$ were to have two components K and L, then there could not be two independent paths in G joining the given vertices in K and L, respectively. Hence $H' - \{t\}$ is connected for each $t \in VG$. It also follows that $G - \{e\}$ and $G - \{a\}$ are connected and therefore so are $H' - \{u\}$ and $H' - \{v\}$. From Theorem 1.25, we conclude that H' is 2-connected, and therefore so is H.

We now infer that at least two independent paths in H join vertices u and v. The union of these paths is a circuit D. The observation that $(D \cap EG) \cup \{e, a\}$ is a circuit in G completes the proof. ■

1.14 Blocks

Let G be a graph. A *block* of G is a maximal connected subgraph with the property that any two distinct edges belong to a common circuit. Thus if e is a loop or isthmus of G, then $G[\{e\}]$ is a block of G. All other blocks have more than one edge. Moreover, any block with more than two vertices has a skeleton with at least two edges, and this skeleton is 2-connected by Theorem 1.26. Such blocks are therefore 2-connected. It follows that any block in G must be 2-connected unless its edge set contains only an isthmus of G.

We insert some further results about blocks at this point.

Theorem 1.27 *The edge sets of the blocks of a graph G partition EG.*

Proof. Choose an edge $e \in EG$, and suppose that e were to belong to blocks K and L. We shall prove that any two distinct edges of $EK \cup EL$ belong to a common circuit.

Choose distinct edges $a, b \in EK \cup EL$. If $a = e$ or $b = e$, then by the definition of a block there is a circuit included in $EK \cup EL$ containing a and b. Suppose therefore that $e \notin \{a, b\}$. Then there exists a circuit $C \subseteq EK \cup EL$ containing e and a and another circuit $D \subseteq EK \cup EL$ containing e and b. If C does not contain b, then let P be the path in the graph $G[C \cup D]$ that is of maximal length with respect to the property of containing b and having no internal vertex in VC. P therefore joins two vertices, u and v, of VC. There is a path Q, included in C, which joins u to v and contains a. Then $P \cup Q$ is a circuit in $G[EK \cup EL]$ that contains a and b. The maximality of K and L now reveals that both must equal $G[EK \cup EL]$. ■

Theorem 1.28 *Let K and L be distinct blocks of a graph G. Then*

$$|VK \cap VL| \leq 1.$$

Proof. Suppose that $|VK \cap VL| > 1$. Then there exists a path P in K joining two vertices u and v of VL and having no internal vertex in VL. Likewise there is a path Q in L joining u and v, and $P \cap Q = \emptyset$ by Theorem 1.27. Hence $P \cup Q$ is a circuit that meets both EK and EL. This conclusion violates Theorem 1.27. ■

Theorem 1.29 *A graph G has a block L that contains at most one vertex of $V[G - EL]$.*

Proof. Suppose the contrary. Let B be the set of blocks of G, and let A be the set of vertices of G that belong to more than one block. Let H be a simple bipartite graph with bipartition $\{A, B\}$, where a vertex $u \in A$ is adjacent to a vertex $L \in B$ if and only if $u \in VL$. The degree of each vertex of H is then at least 2. (Observe that if $L \in B$, then every vertex of $V[G - EL]$ that is in L also belongs to another block of G and is therefore a member of A adjacent to L in H.) Hence H has a circuit C by Corollary 1.19. Let $L_1, L_2, \ldots, L_n$ be the blocks of G in VC, and define $K = G[\bigcup_{i=2}^{n} EL_i]$. Corollary 1.5 guarantees the connectedness of K. Hence, there is a path P in K, of minimal length, joining two distinct vertices of VL_1. Let x and y be the vertices joined by P. There exists a path Q in L_1 joining x and y, and $P \cup Q$ must be a circuit D by the choice of P. However, not every edge of D belongs to L_1, in contradiction to Theorem 1.27. ■

1.15 The cycle and cocycle spaces of a graph

We now introduce some vector spaces defined on graphs. Note first that if S is any set, then $\mathcal{P}(S)$ defines a vector space of dimension $|S|$ over the field $\mathcal{Z}_2$ of residue classes modulo 2, with the sum of sets defined in the usual way. We may define the inner product of sets X and Y to be 0 if $|X \cap Y|$ is even and 1 otherwise. We take $S = EG$ for some graph G, and consider some subspaces of $\mathcal{P}(EG)$.

Theorem 1.30 *For any graph G, the collection of the edge sets of the Eulerian subgraphs of G defines a vector space.*

Proof. Let $\mathcal{C}(G)$ be the collection of the edge sets of the Eulerian subgraphs of G. Since $\emptyset \in \mathcal{C}(G)$, it suffices to establish the closure of $\mathcal{C}(G)$ under addition. Choose $X \in \mathcal{C}(G)$ and $Y \in \mathcal{C}(G)$. Then $G[X]$ and $G[Y]$ are Eulerian subgraphs of G. It is immediate that $G[X + Y]$ is Eulerian, and the theorem follows. ■

We call $\mathcal{C}(G)$ the *cycle space* of G. Its elements are *cycles*. Thus, each cycle is a sum of disjoint circuits. The orthogonal complement, $\mathcal{C}^{\perp}(G)$, of $\mathcal{C}(G)$ is the *cocycle space* of G.

Theorem 1.31 *The elements of the cocycle space of a graph G are the cocycles of G.*

Proof. The circuits of G span $\mathcal{C}(G)$. Thus, if X is a cocycle of G, then $|X \cap C|$ is even for every cycle C in G by Theorem 1.23 (since the operation of intersection is distributive over addition). Hence $X \in \mathcal{C}^{\perp}(G)$. Conversely, if $X \in \mathcal{C}^{\perp}(G)$, then $|X \cap C|$ is even for every circuit C, and X is a cocycle by Theorem 1.23. ■

Theorem 1.32 *For any graph G,*

$$\dim \mathcal{C}^{\perp}(G) = |VG| - c(G),$$

and $\mathcal{C}^{\perp}(G)$ is spanned by the vertex cocycles of G.

Proof. Let $H_1, H_2, \ldots, H_k$ be the components of G, and let $VG = \{v_1, v_2, \ldots, v_m\}$. Without loss of generality we may assume that $v_i \in VH_i$ for all $i \in J_k$. Let $B = \{v_{k+1}, v_{k+2}, \ldots, v_m\}$, and define $\mathcal{B} = \{\partial v_{k+1}, \partial v_{k+2}, \ldots, \partial v_m\}$. Thus $\mathcal{B}$ is a set of vertex cocycles.

We show first that $\mathcal{B}$ is linearly independent. Any sum of elements of $\mathcal{B}$ is of the form ∂S for some $S \subseteq B$. Suppose therefore that $\partial S = \emptyset$, but that $S \neq \emptyset$. Let T be a minimal non-empty subset of S such that $\partial T = \emptyset$. Since $\partial U \neq \emptyset$ for all non-empty proper subsets U of T, it follows by definition that $G[T]$ is a component H_j of G. But by construction $VH_j \not\subseteq B$, in contradiction to the fact that $S \subseteq B$. Thus, if $\partial S = \emptyset$, then $S = \emptyset$. Hence $\mathcal{B}$ is linearly independent.

Next, for any $i \in J_k$ we have $\partial v_i = \partial(VH_i - \{v_i\})$, which is a sum of cocycles in $\mathcal{B}$. Hence $\mathcal{B}$ spans the set of vertex cocycles in G. Since any cocycle is a sum of vertex cocycles, it follows that $\mathcal{B}$ spans $\mathcal{C}^{\perp}(G)$. Hence $\dim \mathcal{C}^{\perp}(G) = |B| = m - k = |VG| - c(G)$. ■

Corollary 1.33 *For any graph G,*

$$\dim \mathcal{C}(G) = |EG| - |VG| + c(G).$$

Proof. Since $\dim \mathcal{C}(G) + \dim \mathcal{C}^{\perp}(G) = \dim \mathcal{P}(EG) = |EG|$, it follows that $\dim \mathcal{C}(G) = |EG| - \dim \mathcal{C}^{\perp}(G) = |EG| - |VG| + c(G)$. ■

Corollary 1.34 *For any graph G, $\dim \mathcal{C}(G)$ is the sum of the dimensions of the cycle spaces of the blocks of G.*

Proof. Clearly, we may assume that G is connected. Since G and $G - \{e\}$ have the same cycle space for any isthmus e, we may also assume that G has no isthmus. The corollary is also obvious if G has only one block. We may therefore assume that G has more than one block and that the result holds for any graph with fewer blocks than G. By Theorem 1.29 and the connectedness of G, we may choose a block L that contains a unique vertex v of $V[G - EL]$. Let K be the component of $G - EL$ that contains v. Then

$$\begin{aligned}
\dim \mathcal{C}(G) &= |EG| - |VG| + 1 \\
&= |EL| + |EK| - (|VL| + |VK| - 1) + 1 \\
&= |EL| - |VL| + 1 + |EK| - |VK| + 1 \\
&= \dim \mathcal{C}(L) + \dim \mathcal{C}(K),
\end{aligned}$$

as required. ■

1.16 Double covers

A family $(C_1, C_2, \ldots, C_n)$ of non-empty cycles in a graph G is a *cycle double cover* of G if, for each edge $e \in EG$, e belongs to C_i for exactly two values of $i \in J_n$.

A family of cycles in a graph G is said to be a *spanning family* if its components span $\mathcal{C}(G)$. Similarly, we may talk about a *spanning set* of cycles of G. If the cycles of the spanning family or spanning set constitute a cycle double cover, then this double cover is also described as *spanning*. Similarly, we also talk about a *spanning circuit double cover*. We insert some results about cycle double covers.

Lemma 1.35 *Let $(C_1, \ldots, C_n)$ be a cycle double cover $\mathcal{C}$ of a connected graph G, where $n > 0$. Let $\mathcal{D}$ be the subspace of $\mathcal{C}(G)$ spanned by the components of $\mathcal{C}$. Suppose that each pair of adjacent edges of G is a subset of a circuit in $\mathcal{D}$. Then* $\dim \mathcal{D} = n - 1$.

Proof. Since $\sum_{i=1}^{n} C_i = \emptyset$, we have $\dim \mathcal{D} < n$. On the other hand, suppose without loss of generality that $\sum_{i=1}^{j} C_i = \emptyset$ for some $j \in J_{n-1}$. Let $T = \bigcup_{i=1}^{j} C_i$. Since $j > 0$, we have $T \neq \emptyset$. Moreover, $EG - T \neq \emptyset$, for otherwise each edge of G would belong to C_i for exactly two choices of $i \leq j$ since $\sum_{i=1}^{j} C_i = \emptyset$, and so we should have the contradiction that $j = n$. As G is connected, Corollary 1.9 shows that there is a vertex v incident on an edge $e \in T$ and on an edge $f \in EG - T$. By hypothesis, these edges are links contained in some circuit $C \in \mathcal{D}$. Now, by the definition of T we see that $|\partial v \cap T \cap C_i|$ is even for each i, yet $|\partial v \cap T \cap C| = 1$. These results contradict the hypothesis that the components of $\mathcal{C}$ span $\mathcal{D}$. Hence, the cycles $C_1, \ldots, C_{n-1}$ are independent, and the lemma follows. ■

Remark: The proof of this lemma shows that any $n-1$ components of $\mathcal{C}$ span $\mathcal{D}$.

Corollary 1.36 *Let* $(C_1, \ldots, C_n)$ *be a spanning cycle double cover of a 2-connected graph G. Then* $\dim \mathcal{C}(G) = n - 1$.

Proof. Since G is 2-connected, Theorem 1.26 shows that the hypotheses of Lemma 1.35 are satisfied with $\mathcal{D} = \mathcal{C}(G)$. The result follows. ■

Corollary 1.37 *Any spanning cycle double cover of a 2-connected graph is a circuit double cover.*

Proof. Let $(C_1, \ldots, C_n)$ be a spanning cycle double cover of a 2-connected graph G. Each cycle C_i is the sum of a set S_i of disjoint circuits. It follows from Corollary 1.36 that $\sum_{i=1}^{n} |S_i| = \dim \mathcal{C}(G) + 1 = n$. Hence $|S_i| = 1$ for each i, and each C_i is a circuit. ■

Lemma 1.38 *Let* $(C_1, \ldots, C_n)$ *be a spanning circuit double cover of a 2-connected graph G. If G has a vertex of degree greater than 2, then* $C_1, \ldots, C_n$ *are distinct.*

Proof. Suppose without loss of generality that $C_1 = C_2$. Since G is connected, we may choose a vertex $v \in VC_1$ of degree at least 3 in G. Let e and a be the edges of C_1 incident on v. Then $C_i \cap \{e, a\} = \emptyset$ for each $i > 2$, since $(C_1, \ldots, C_n)$ is a circuit double cover. There exists a third edge b incident on v. Since G is 2-connected, there is a circuit C in G passing through a and b. Hence $e \notin C$. Thus $|C \cap \{e, a\}| = 1$, though $|C_i \cap \{e, a\}|$ is even for each i. We thus have a contradiction to the hypothesis that $C_1, \ldots, C_n$ span $\mathcal{C}(G)$. ■

2
Maps

The introduction to maps given here follows the treatment in Tutte (1973, 1984). Its relationship to other approaches to this subject will be investigated later. One of these approaches, via 3-graphs, is the route that will be followed in subsequent chapters, but we first need to reconcile it with the more traditional development outlined here.

2.1 Permutations

A *permutation* of a set X is a bijection φ from X onto itself. The image of an object $x \in X$ under φ is denoted by φx. Given a permutation φ of X, it is clear that for each $x \in X$ there exists a smallest positive integer r satisfying $\varphi^r x = x$. Let $Y = \{\varphi x, \varphi^2 x, \ldots, \varphi^r x\}$. Then Y is a *cell* or *orbit* of φ. Moreover, the restriction, $\varphi|_Y$, of φ to Y is a *cycle* of φ of *length* r and will be denoted by $(\varphi x\ \varphi^2 x \cdots \varphi^r x)$. We observe that the cells of φ form the cells of a partition of X. This partition is said to be *induced* by φ. A *cyclic* permutation is one that induces a trivial partition. Note that if permutations φ and φ' both induce refinements of a partition $\mathcal{P}$, then so does $\varphi\varphi'$, for if $p \in P \in \mathcal{P}$, then $\varphi' p \in P$ and so $\varphi\varphi' p \in P$.

Let $\varphi_1, \varphi_2, \ldots, \varphi_m$ be permutations. Let $\mathcal{D} = \bigcup_{i=1}^{m} \mathcal{D}\varphi_i$. For each i, let ψ_i denote the permutation of $\mathcal{D}$ such that $\psi_i x = \varphi_i x$ for all $x \in \mathcal{D}\varphi_i$ and $\psi_i x = x$ for all $x \notin \mathcal{D}\varphi_i$. Then the *product*, $\prod_{i=1}^{m} \varphi_i$, of $\varphi_1, \varphi_2, \ldots, \varphi_m$ is defined as $\prod_{i=1}^{m} \psi_i$. In this way, we may describe a permutation as the product of its cycles.

The identity permutation is written as id. A permutation φ is an *involution* if $\varphi^2 = \text{id}$.

2.2 Maps

A map is often described topologically as a cellular decomposition of a surface. Our commitment to a combinatorial approach precludes adoption of this definition, but we replace it by one that should make the topological connection evident.

The idea of using permutations to define maps combinatorially appeared in Edmonds (1960). The concept was pursued further by Jacques (1968) and Walsh and Lehman (1972), and their work forms the foundation of the development we will follow. Another treatment along similar lines has been given by Tutte (1984), and will be discussed later.

Let X be a set with even cardinality. Its elements are called *ends*. Let π be a permutation of X that induces a partition EM of X in which every cell is a pair. Let φ be a second permutation of X, and let VM be the partition of X induced by it. Then (X, π, φ) is a *map*, M, and VM and EM, respectively, are the vertex set and edge set of a graph G in which each edge joins the vertices it meets. Thus $VG = VM$ and $EG = EM$. We call VM and EM the *vertex* set and *edge* set, respectively, of M. Note that $X = \bigcup EG = \bigcup VG$. M is said to *contain* the elements of $VM \cup EM$. Let Z be any set whose elements are sets disjoint from X, and let G^* be the graph with vertex set $VG \cup Z$ and edge set EG such that G is a subgraph of G^*. (Thus the vertices of $VG^* - VG$ are isolated.) Then G^* is a graph *underlying* M. However, when we speak of *the* graph underlying M, we mean G, the one with no isolated vertices.

In Chapter 1, we saw how a graph may be represented by a drawing in the plane. Suppose we are given a drawing that represents a graph G underlying a map (X, π, φ). It is intuitively clear that if G has a vertex of degree greater than 2, then the drawing induces two permutations of X whose cells are the vertices of G, these permutations being inverses of each other. The drawing may be chosen so that φ is one of the induced permutations. It is therefore more accurate to think of the drawing as representing a map that G underlies. However, the same drawing may represent different maps. For instance, if σ is a permutation of X that induces a refinement of the partition induced by π, then (X, π, φ) and $(X, \pi, \sigma\varphi\sigma)$ can be represented by a common drawing.

Example 2.1 *Let*

$$\begin{aligned} X &= \{e_1^+, e_1^-, e_2^+, e_2^-, e_3^+, e_3^-, e_4^+, e_4^-\}, \\ \pi &= (e_1^+ e_1^-)(e_2^+ e_2^-)(e_3^+ e_3^-)(e_4^+ e_4^-), \\ \varphi &= (e_1^+ e_2^- e_1^- e_3^+)(e_2^+ e_3^- e_4^+), \end{aligned}$$

and $\mathcal{D}\varphi = \mathcal{D}\pi = X$. Thus (X, π, φ) is a map, M. Moreover, the set EM of cells of π is $\{e_1, e_2, e_3, e_4\}$, where $e_i = \{e_i^+, e_i^-\}$ for all $i \in J_4$, and the set VM of cells of φ is $\{v_1, v_2, v_3\}$, where $v_1 = \{e_1^+, e_1^-, e_2^-, e_3^+\}$, $v_2 = \{e_2^+, e_3^-, e_4^+\}$, and $v_3 = \{e_4^-\}$. Let $Z = \{\{X\}, \{\{X\}\}\}$. Since $X \notin X$ and $\{X\} \notin X$, $VM \cup Z$ and EM are the vertex set and edge set, respectively, of a graph G underlying M. M is represented by Figure 2.1, but the same drawing (apart from the labels) also represents, for

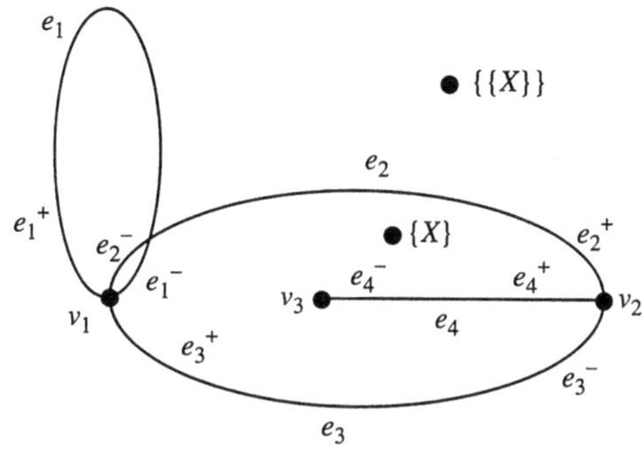

FIGURE 2.1.

example, the map (X, π, φ'), where $\varphi' = (e_1^- e_2^- e_1^+ e_3^-)(e_2^+ e_3^+ e_4^+)$ and $\mathcal{D}\varphi' = X$. Note that $\varphi' = \sigma\varphi\sigma$, where $\sigma = (e_1^+ e_1^-)(e_3^+ e_3^-)$, and σ induces a refinement of EM. Since σ^2 is the identity permutation, we also have $\varphi = \sigma\varphi'\sigma$.

Any graph is isomorphic to one that underlies a map. To see this, it suffices to consider graphs with no isolated vertices. Let G be such a graph, with $VG = \{v_1, v_2, \ldots, v_m\}$ and $EG = \{e_1, e_2, \ldots, e_n\}$. For each $i \in J_n$, define $e_i^+ = (e_i, 1)$, $e_i^- = (e_i, -1)$, and $e_i' = \{e_i^+, e_i^-\}$, and write $X = \bigcup_{i=1}^n e_i' = EG \times \{-1, 1\}$. Let $EG' = \{e_1', e_2', \ldots, e_n'\}$. Let $\{v_1', v_2', \ldots, v_m'\}$ be a partition VG' of X with the property that, for each $(i, j, k) \in J_n \times J_m \times J_m$ such that e_i joins v_j to v_k in G, we have $(v_j' \cap e_i') \cup (v_k' \cap e_i') = e_i'$, $v_j' \cap e_i' \neq \emptyset$ and $v_k' \cap e_i' \neq \emptyset$. Then VG' and EG' are the vertex set and edge set, respectively, of a graph G' isomorphic to G and underlying a map.

Maps (X, π, φ) and (X', π', φ') are *isomorphic* if there exists a bijection $f : X \to X'$ such that $f\pi = \pi' f$ and $f\varphi = \varphi' f$. We say that f is an *isomorphism* from (X, π, φ) to (X', π', φ'). Isomorphism of maps also defines an equivalence relation, and one usually considers an equivalence class of isomorphic maps. The set of all maps isomorphic to a map M will be denoted by $[M]$, or simply M.

2.3 Imbeddings of maps

Let (X, π, φ) be a map M with edge set EM, and let $\iota : EM \to \{-1, 1\}$ be a function. Then (M, ι) is an *imbedding* M^ι of M and of any graph G underlying M. Thus $M^\iota = (X, \pi, \varphi, \iota)$. Imbeddings (M, ι) and (M', ι') are *isomorphic* if there is an isomorphism f from M to M' such that $\iota(\{x, \pi x\}) = \iota'(\{f(x), f(\pi x)\})$ for all $x \in X$. The set of all imbeddings isomorphic to (M, ι) will be denoted by $[M, \iota]$ or (M, ι). An edge e is *barred* if $\iota(e) = -1$, and *unbarred* otherwise. We also define $\iota(x) = \iota(e)$ for each end $x \in e$. M^ι is an *oriented* imbedding if every edge is unbarred. An oriented imbedding (X, π, φ, ι) will also be written as (X, π, φ). Given a drawing that represents M, we may represent M^ι by placing a mark, such as a transverse bar, on each curve that represents a barred edge.

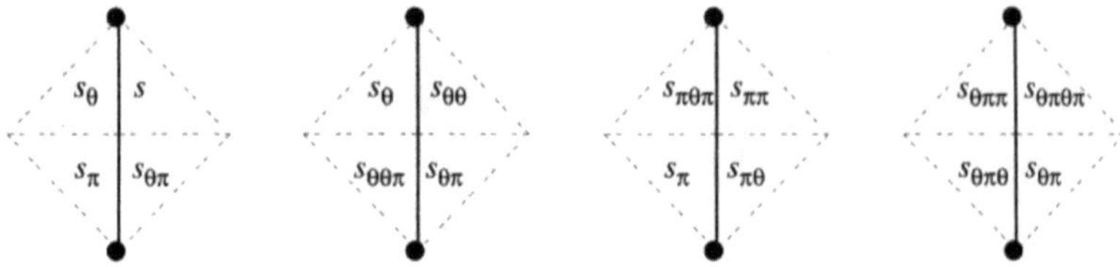

FIGURE 2.2.

We now attempt to motivate our definition of an imbedding by developing the idea of a face.

Let (X, π, φ) be a map M with edge set EM. For any $x \in X$, the ordered pairs $(x, 1)$ and $(x, -1)$ are the *sides* of x. For each $x \in X$, we define $s(x) = (x, 1)$, $s_\theta(x) = (x, -1)$, $s_\pi(x) = s(\pi x)$, and $s_{\theta\pi}(x) = s_\theta(\pi x)$. We may now define $s_{\theta\theta} = s_{\pi\pi} = s_{\theta\pi\theta\pi} = s$, $s_{\pi\theta\pi} = s_{\theta\pi\pi} = s_\theta$, $s_{\theta\theta\pi} = s_{\theta\pi\theta} = s_\pi$, and $s_{\pi\theta} = s_{\theta\pi}$. If $x \in e \in EM$, we may think intuitively of $s(x)$ and $s_\theta(x)$ as the sides of x in the geometric sense, and of $s_\pi(x)$ and $s_{\theta\pi}(x)$ as the sides of the opposite end of e. The application of the same intuition to the functions s_θ, s_π, and $s_{\theta\pi}$ motivates the above equations. (See Figure 2.2, where the sides of x and πx are represented by triangles.) The pairs $\{s(x), s_\theta(x)\}$ and $\{s_\pi(x), s_{\theta\pi}(x)\}$ are described as *transverse*, $\{s(x), s_{\theta\pi}(x)\}$ and $\{s_\theta(x), s_\pi(x)\}$ as *longitudinal*, and $\{s(x), s_\pi(x)\}$ and $\{s_\theta(x), s_{\theta\pi}(x)\}$ as *diagonal*. These six pairs are called the *pairs* of x.

We now introduce some functions that act on sides of ends. First, given an imbedding (X, π, φ, ι) of a map M, define

$$\Pi(x, k) = (\varphi^{k\iota(x)}\pi x, k\iota(x))$$

for each $(x, k) \in X \times \{-1, 1\}$. [Thus $\varphi^{k\iota(x)} = \varphi^{-1}$ if $k\iota(x) = -1$ and $\varphi^{k\iota(x)} = \varphi$ otherwise. See Figure 2.3.] Letting $x' = \pi\varphi^{-k}x$ and $k' = k\iota(\varphi^{-k}x)$, we find that

$$\begin{aligned} k'\iota(x') &= k\iota(\varphi^{-k}x)\iota(\pi\varphi^{-k}x) \\ &= k\iota^2(\varphi^{-k}x) \\ &= k, \end{aligned}$$

and so

$$\begin{aligned} \varphi^{k'\iota(x')}\pi x' &= \varphi^k\pi^2\varphi^{-k}x \\ &= x; \end{aligned}$$

hence $\Pi(x', k') = (x, k)$. Thus, Π maps the set $X \times \{-1, 1\}$ onto itself, and Π is therefore a permutation of $X \times \{-1, 1\}$. Note also that

$$\Pi^{-1}(x, k) = (x', k') = (\pi\varphi^{-k}x, k\iota(\varphi^{-k}x)).$$

The cells of Π are the *orbits* of M^ι. If M^ι is oriented, then the *orbits* of M are those of M^ι.

We now require some lemmas.

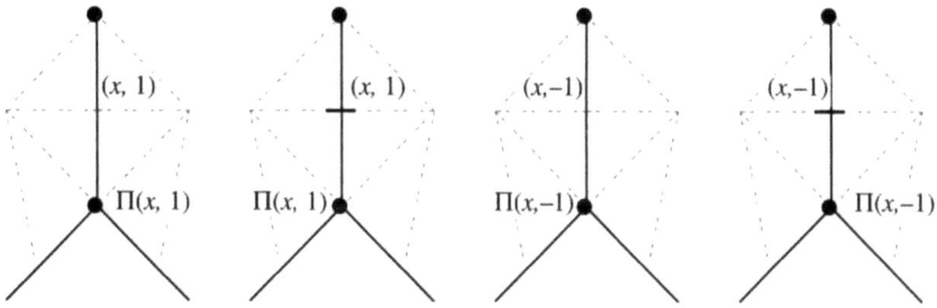

FIGURE 2.3.

Lemma 2.1 *For any $x \in X$ and $k \in \{-1, 1\}$, (x, k) and $(\pi x, -k\iota(x))$ belong to distinct orbits of M^ι.*

Proof. If not, there exists $r > 0$ for which $\Pi^r(x, k) = (\pi x, -k\iota(x))$. Let us suppose that the triple (x, k, r) is chosen to minimise r. Clearly $r > 1$, for otherwise $-k\iota(x) = q(\Pi(x, k)) = k\iota(x)$. If $r = 2$, then

$$\Pi(\varphi^{k\iota(x)}\pi x, k\iota(x)) = (\pi x, -k\iota(x)),$$

so that

$$\begin{aligned} \varphi^{k\iota(x)}\pi x &= p(\Pi^{-1}(\pi x, -k\iota(x))) \\ &= \pi\varphi^{k\iota(x)}\pi x, \end{aligned}$$

another contradiction. Hence $r > 2$. Since $\Pi(x, k) = (\varphi^{k\iota(x)}\pi x, k\iota(x))$, we have

$$\begin{aligned} \Pi^{r-2}\Pi(x, k) &= \Pi^{-1}\Pi^r(x, k) \\ &= \Pi^{-1}(\pi x, -k\iota(x)) \\ &= (\pi\varphi^{k\iota(x)}\pi x, -k\iota(x)\iota(\varphi^{k\iota(x)}\pi x)) \\ &= (\pi p, -q\iota(p)), \end{aligned}$$

where $p = p(\Pi(x, k)) = \varphi^{k\iota(x)}\pi x$ and $q = q(\Pi(x, k)) = k\iota(x)$. Therefore, the initial choice of (x, k, r) is contradicted by the triple $(\Pi(x, k), r - 2)$, since $r > 2$. ■

For each $(x, k) \in X \times \{-1, 1\}$, we write $\Phi(x, k) = (\pi x, -k\iota(x))$. (See Figure 2.4.)

Lemma 2.2 *Let P be the orbit of M^ι that contains (x, k) and Q the orbit that contains $\Phi(x, k)$. If $(y, l) \in P$, then $\Phi(y, l) \in Q$.*

Proof. It suffices to establish the result for $(y, l) = \Pi(x, k)$ and appeal to induction. With $y = p(\Pi(x, k))$ and $l = q(\Pi(x, k))$, we have

$$\begin{aligned} \Pi^{-1}(\Phi(x, k)) &= \Pi^{-1}(\pi x, -k\iota(x)) \\ &= (\pi y, -l\iota(y)) \\ &= \Phi(y, l), \end{aligned}$$

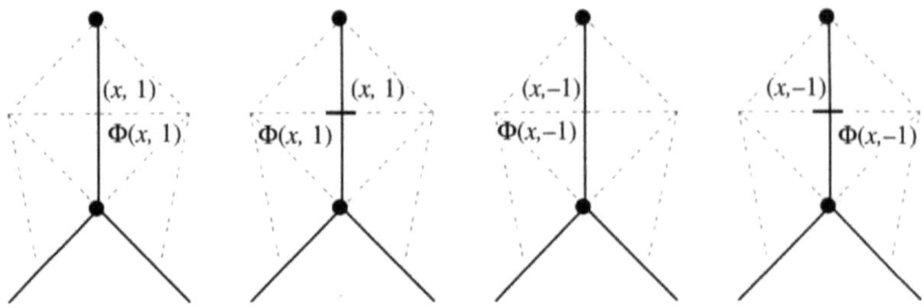

FIGURE 2.4.

and so $\Phi(y, l) \in Q$. ■

We denote Q by P^-. Lemma 2.2 shows that P^- is well defined. Note that Q^- contains $\Phi^2(x, k) = (\pi^2 x, k\iota(x)\iota(\pi x))$. Since π^2 is the identity permutation of X and $\iota(\pi x) = \iota(x)$, it follows that $(x, k) \in Q^-$, and so $Q^- = P$. The set of all pairs of the form $\{z, \Phi(z)\}$, where $z \in P$, is a *face* of M^ι, the face *determined* by P (or P^-). If M^ι is oriented, then the *faces* of M are those of M^ι. Observe that if the pair $\{t(x), u(\pi x)\}$ belongs to a face f, where $t(x)$ and $u(\pi x)$ are sides of ends x and πx, respectively, then $\{t_\theta(x), u_\theta(\pi x)\}$ also belongs to a face, which may be f. [*Proof*: Letting $t(x) = (x, k)$ so that $u(\pi x) = (\pi x, -k\iota(x))$, we have $t_\theta(x) = (x, -k)$; hence $\Phi(t_\theta(x)) = (\pi x, k\iota(x)) = u_\theta(\pi x)$.]

Let G be a graph underlying a map M with an imbedding I. Then we write $VI = VM = VG$ and $EI = EM = EG$. These sets are, respectively, the *vertex set* and *edge set* not only of G and M but also of I. Similarly, their elements are, respectively, the *vertices* and *edges* of M and I. Moreover, we write FI for the set of faces of I, and we write $FM = FI$ if I is oriented.

Each $x \in X$ is an element of a unique vertex and a unique edge; these we denote by v_x and e_x, respectively. The set

$$S_x = \{(x, -1), (x, 1), (\pi x, -1), (\pi x, 1)\}$$

can be partitioned into pairs that are elements of faces of M^ι. We say that x, v_x, and e_x *belong* to these faces. Thus x and e_x belong to, at most, two distinct faces. If e_x belongs to two distinct faces, it is *bifacial* in I; otherwise, it is *monofacial* in I. If e_x is monofacial in I, we may choose an orbit P of M^ι that contains two elements of S_x. Then P^- contains the other two elements of S_x. If $|P \cap \{(x, 1), (x, -1)\}| = 1$, then $|P^- \cap \{(x, 1), (x, -1)\}| = 1$ and we say that e_x is *coherent* in I. (See Figure 2.5.) An edge that is monofacial in I but not coherent in I is *consistent* in I. (See Figure 2.6.)

Next, let f be the face $\{\{z_1, \Phi(z_1)\}, \{z_2, \Phi(z_2)\}, \ldots, \{z_m, \Phi(z_m)\}\}$. Without loss of generality, we may assume that $\Pi(z_i) = z_{i+1}$ for all $i \in J_{m-1}$, and $\Pi(z_m) = z_1$. For each $i \in J_m$, let $p(z_i) = x_i$, so that $x_i \in X$, and let $x_i \in e_i \in EI$. Note that $p(\Phi(z_i)) = \pi p(z_i) = \pi x_i \in e_i$, and therefore e_i is well defined. It also follows from the definition of $\Pi(z_i)$ that for each $i \in J_{m-1}$ the edges e_i and e_{i+1} are adjacent or equal (and of course the same is true of e_m and e_1). The sets $[e_1 e_2 \cdots e_m]$ and $[e_m e_{m-1} \cdots e_1]$ are the *oriented boundaries* of f. A capital letter appearing as part

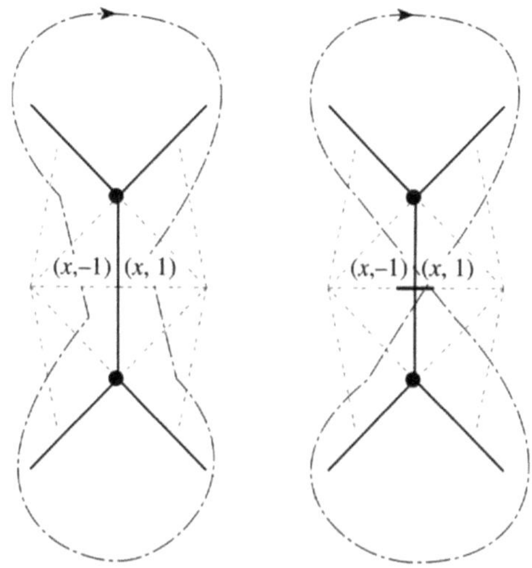

FIGURE 2.5.

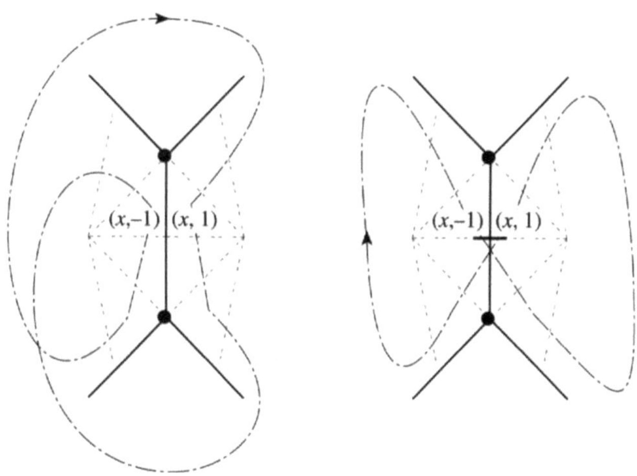

FIGURE 2.6.

of the notation for an oriented boundary is to be interpreted as representing a string of symbols. We say that f is *determined* by either of its oriented boundaries. Note that if a single oriented boundary is selected for each face, then any edge e appears exactly twice in the resulting list of oriented boundaries. These appearances are in distinct oriented boundaries if and only if e is bifacial. The collection of all bifacial edges in $\{e_1, e_2, \ldots, e_m\}$ is the *boundary*, ∂f, of f. It follows that ∂f is a cycle. The *boundary*, ∂S, of a set S of faces is the sum of the boundaries of the faces in S. Thus $\partial\{f\} = \partial f$. Note that ∂S is also a cycle.

Let $[e_1 e_2 \cdots e_m]$ be an oriented boundary of a face f, and suppose that e_1 is monofacial. Then there exists a unique integer $j > 1$ such that $e_1 = e_j$. The set of

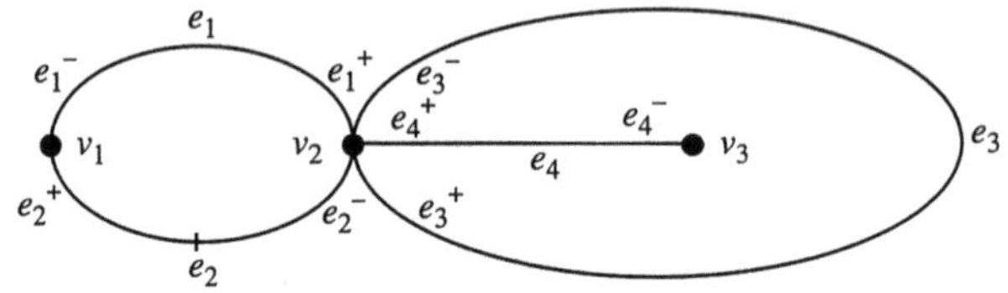

FIGURE 2.7.

all bifacial edges in $\{e_2, e_3, \ldots, e_{j-1}\}$ is called a *boundary segment* of f in I with respect to e_1, as is the set of all bifacial edges in $\{e_{j+1}, e_{j+2}, \ldots, e_m\}$.

Example 2.2 *Let (X, π, φ, ι) be the imbedding I drawn in Figure 2.7. Here, $X = \{e_1^+, e_1^-, e_2^+, e_2^-, e_3^+, e_3^-, e_4^+, e_4^-\}$, $e_i = \{e_i^+, e_i^-\}$ for all $i \in J_4$, $EI = \{e_1, e_2, e_3, e_4\}$, $\varphi = (e_1^- e_2^+)(e_2^- e_1^+ e_3^- e_4^+ e_3^+)$, $\iota(e_1) = \iota(e_3) = \iota(e_4) = 1$, and $\iota(e_2) = -1$. This imbedding has just two faces, and they are determined by the oriented boundaries $[e_1e_2e_1e_2e_3]$ and $[e_3e_4e_4]$, respectively. The only bifacial edge is e_3; e_1 and e_2 are consistent, whereas e_4 is coherent. Both faces have $\{e_3\}$ as a boundary.*

2.4 3-graphs

We now return to graphs. A graph is *regular* of *degree n*, or *n-valent*, if the degree of every vertex is n. Thus, the graphs induced by circuits are the non-null connected 2-valent graphs. A 3-valent graph is *cubic*. The edge set of a 1-valent spanning subgraph of a graph G is a *1-factor* of G. Observe that the sum of two 1-factors is a union of disjoint circuits of even length.

Let H be a non-null cubic graph. It may be that H has two disjoint 1-factors, V and F. If so, then $EH + V + F$ is a third 1-factor, E. Moreover, H has no loops. We say that the ordered triple (H, V, F) is a *3-graph*. The 1-factors V, F, and E are called *colours*, and their elements are *blue*, *red*, and *yellow*, respectively. Circuits included in $V + F$, $V + E$, or in $F + E$ are called *bicoloured polygons* or *bigons* of H and of the 3-graph. They may be identified as *red-blue* bigons, *blue-yellow* bigons, and *red-yellow* bigons, respectively.

Example 2.3 *The 3-graph illustrated in Figure 2.8 has 4 red-blue bigons, 2 blue-yellow bigons, and 3 red-yellow bigons, one of which is a digon.*

By abuse of language, properties of H may be ascribed to the 3-graph (H, V, F).

A connected 3-graph is a special case of what Vince (1983) calls a *combinatorial map*. A combinatorial map is simply a connected r-valent graph whose edges have been coloured in r colours so that edges incident on a common vertex are coloured differently. Combinatorial maps generalise not only the maps we are studying but also polytopes, tessellations, and higher dimensional analogues of imbeddings. However we press this matter no further and leave it to the reader to seek further details from Vince (1983).

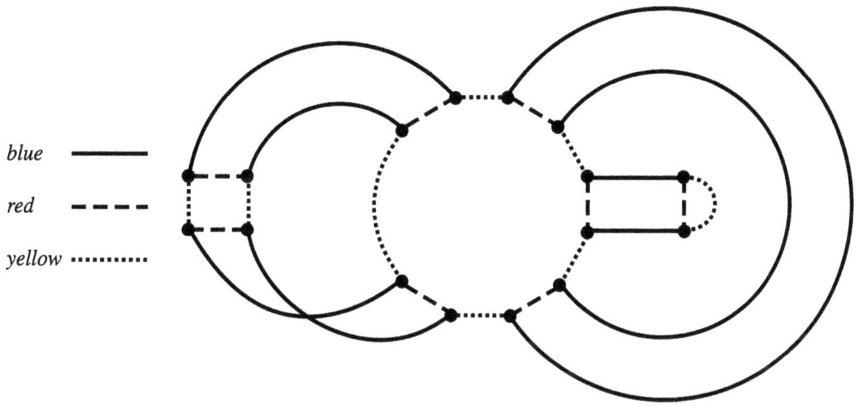

FIGURE 2.8.

It may also be true that the red-blue bigons of a 3-graph are all squares. In this case, the 3-graph is called a *gem*. [This terminology was coined by Lins (1982), and is an acronym for "graph-encoded map".] The red-blue bigons of a gem sometimes are called *bisquares*.

Let K be a gem (H, V, F). Every red-blue bigon has two red edges, each of which belongs to a unique red-yellow bigon. Thus, each bisquare meets just two red-yellow bigons, which may not be distinct. Let $G(K)$ be the graph in which $VG(K)$ is the set of red-yellow bigons, $EG(K)$ is the set of red-blue bigons, and each edge joins the vertices it meets. We describe $G(K)$ as the *graph* of K, or the graph *underlying* K, and K as a gem *corresponding* to $G(K)$.

Example 2.4 *Since all red-blue bigons of the* 3*-graph drawn in Figure 2.8 are circuits of length* 4*, this* 3*-graph is a gem. Figure 2.9 illustrates the graph underlying this gem.*

The motivation for studying gems resides in the fact that a cellular imbedding of a non-empty graph G in a closed surface can be modelled by means of a gem in the following way. First, construct the barycentric subdivision Δ of the imbedding of G and colour each vertex of Δ with blue, yellow, or red according to whether it represents a vertex, edge, or face of the imbedding. Each edge of Δ then joins vertices of distinct colours and may be coloured with the third colour. Let K be the dual of Δ, each edge of K being coloured with the colour of the corresponding edge of Δ. Then each red-blue bigon of the 3-graph K is a square, so that K is a gem. (See Figure 2.10. In this figure, the vertices of G are the solid circles, and the edges are the thin solid lines joining such circles. All of the circles are vertices

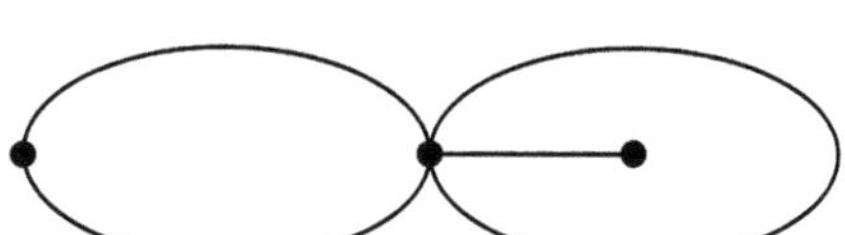

FIGURE 2.9.

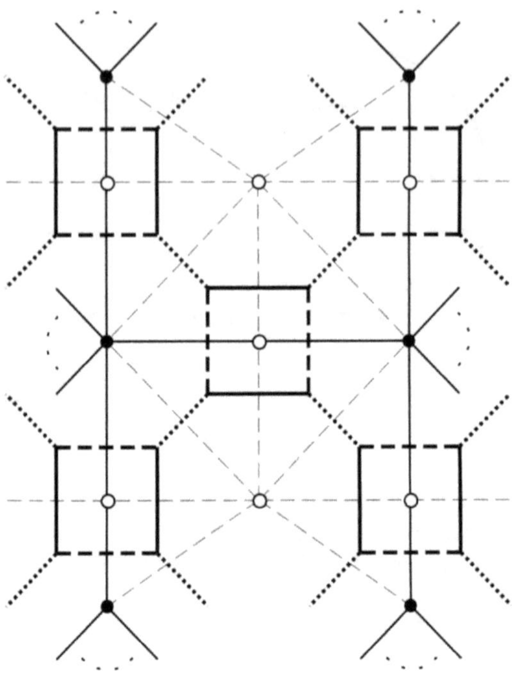

FIGURE 2.10.

of Δ. The edges of Δ are thin solid line segments and the thin dashed lines. The edges of K are thicker and coloured as indicated in the figure. The vertices of K should be self-evident.)

This construction can be reversed. Given K, we first contract each red-yellow bigon to a single vertex. Each red-blue bigon then becomes a digon whose edges are both blue. The identification of the two edges in each of these digons yields G. In fact, $G = G(K)$. Thus, there is a 1:1 correspondence between gems and cellular imbeddings in closed surfaces of graphs whose components are non-empty.

2.5 From maps to gems and back again

We now establish the relationship between maps and gems. Let (X, φ, π, ι) be an imbedding I of a map M. Let $VH = X \times \{-1, 1\}$, and let the elements of VH be vertices of a graph H. For every $x \in X$, let each transverse pair of x be joined in H by an edge coloured red. If $\iota(x) = 1$, let each longitudinal pair of x be joined by an edge coloured blue; if $\iota(x) = -1$, let it be the diagonal pairs of x that are so joined. Finally, let there be an edge coloured yellow joining $(x, -1)$ and $(\varphi x, 1)$. If $X = \emptyset$, then H is the null graph. If $X \neq \emptyset$, then (H, V, F) is a gem, where V and F are the sets of blue and red edges, respectively. This gem is said to *encode* I. For example, the gem in Figure 2.8 encodes the imbedding represented in Figure 2.7. Note that if M is underlaid by a connected graph, then H is connected.

Conversely, let (H, V, F) be a gem K, and let $E = EH + V + F$. Let $X = F$. Let $Y_1, Y_2, \ldots, Y_n$ be the bisquares, and for each $i \in J_n$ let $e_i = Y_i \cap F$. We say that e_i and Y_i *correspond* to each other, and we define $EI = \{e_1, e_2, \ldots, e_n\}$. Clearly, $|e_i| = 2$ for each $i \in J_n$. Let $e_i = \{x_i, x_i'\}$ for each $i \in J_n$, and define $\pi = \prod_{i=1}^{n}(x_i x_i')$. Next, let $B_1, B_2, \ldots, B_m$ be the red-yellow bigons, and for each $i \in J_m$ let $v_i = B_i \cap F$. We say that v_i and B_i *correspond* to each other. Let $VI = \{v_1, v_2, \ldots, v_m\}$. Now choose $v_i = B_i \cap F \in VI$, and let $v_i = \{x_{i1}, x_{i2}, \ldots, x_{it_i}\}$. Let $[x_{i1}y_{i1}x_{i2}y_{i2}\cdots x_{it_i}y_{it_i}]$, where $B_i \cap E = \{y_{i1}, y_{i2}, \ldots, y_{it_i}\}$, be an orbit of B_i. We now define $\varphi_i = (x_{i1}x_{i2}\cdots x_{it_i})$ and $\varphi = \prod_{i=1}^{m}\varphi_i$. Finally, for each $i \in J_m$ and each $j \in J_{t_i}$, let x_{ij} join vertices $(x_{ij}, 1)$ and $(x_{ij}, -1)$ in H, where y_{ij} is incident on $(x_{ij}, -1)$. (See Figure 2.11.) Thus πx_{ij} joins vertices $(\pi x_{ij}, 1)$ and $(\pi x_{ij}, -1)$, where $(\pi x_{ij}, -1)$ is adjacent to $(\varphi\pi x_{ij}, 1)$ in H. Let $e_{ij} = \{x_{ij}, \pi x_{ij}\}$. If there exists an edge of H (coloured blue) joining $(x_{ij}, 1)$ and $(\pi x_{ij}, -1)$, then define $\iota(e_{ij}) = 1$; otherwise, let $\iota(e_{ij}) = -1$. We say that the gem (H, V, F) and the resulting imbedding $I = (X, \pi, \varphi, \iota)$ *correspond*. The vertex set and edge set of I are VI and EI, respectively. An edge of EI is incident on a vertex of VI in $G(K)$ if and only if the corresponding bigons of (H, V, F) meet. Clearly, (H, V, F) encodes I.

Observe that I is not uniquely defined. In the above construction for I, φ_i could be replaced by φ_i^{-1} for any $i \in J_m$ provided that $\iota(e_{ij})$ is also replaced by $-\iota(e_{ij})$ for each link e_{ij} incident on v_i in $G(K)$. Note that this operation preserves the number of faces of the imbedding, and also the set of oriented boundaries of faces.

The observation in the last paragraph motivates the following definition. Let I be an imbedding (X, π, φ, ι) of a map M underlaid by a graph G, and let $S \subseteq VG$. An imbedding $I' = (X, \pi, \varphi', \iota')$ of M is *S-equivalent* to I if we have $\varphi'|_v = \varphi^{-1}|_v$

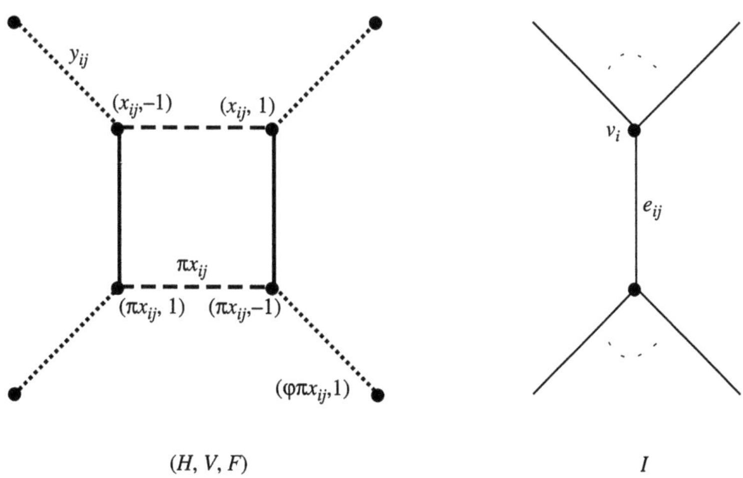

FIGURE 2.11.

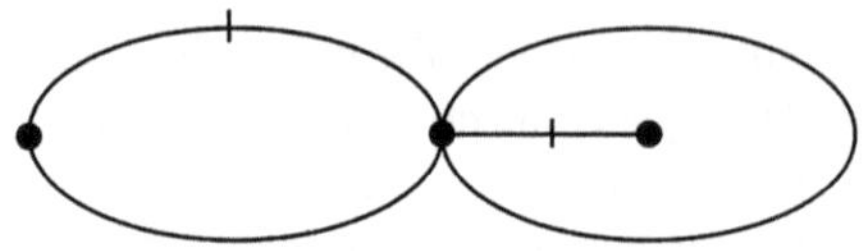

FIGURE 2.12.

for each $v \in S$, $\varphi'|_v = \varphi|_v$ for each $v \notin S$, $\iota'(e) = -\iota(e)$ for each $e = \partial S$, and $\iota'(e) = \iota(e)$ for each $e \notin \partial S$. An imbedding is *equivalent* to I if it is S-equivalent to I for some $S \subseteq VG$. It then corresponds to the same gem as I, and this gem encodes both imbeddings. Note that I is VG-equivalent to $(X, \pi, \varphi^{-1}, \iota)$.

Example 2.5 *Consider the imbedding represented by Figure 2.7 and let G denote the graph that underlies this imbedding. Let S denote the set comprising the vertex of degree* 1 *and the vertex of degree* 2 *in G. Then the imbedding represented in Figure 2.12 is S-equivalent to that represented in Figure 2.7. Furthermore, the gem illustrated in Figure 2.8 encodes them both.*

Let us check that equivalence of imbeddings is in fact an equivalence relation. It is indeed reflexive (take $S = \emptyset$). Clearly, if I' is S-equivalent to I, then I is S-equivalent to I'. Finally, if I' is S-equivalent to I and I'' is T-equivalent to I' (where $T \subseteq VG$), then I'' is $(S+T)$-equivalent to I. Thus, we are entitled to talk about the equivalence class $[I]$ determined by I. We say that any gem that encodes I also *encodes* $[I]$.

Let (H, V, F) be a gem K encoding an imbedding $I = (X, \pi, \varphi, \iota)$. We have already established correspondences between EI and the set of bisquares of K and between VI and the set of red-yellow bigons of K. These correspondences are clearly bijective. We now investigate FI and the set of blue-yellow bigons of K. Let R be such a bigon. Then R contains a blue edge joining a vertex $(x, k) = X \times \{-1, 1\}$ to the vertex $(\pi x, -k\iota(x)) = \Phi(x, k)$, and a yellow edge joining $\Phi(x, k)$ to $(\varphi^{k\iota(x)}\pi x, k\iota(x)) = \Pi(x, k)$. It follows that the set of vertices of VR that are an even distance from (x, k) in $G[R]$ forms a cell P of Π that contains (x, k). The remaining vertices of VR constitute a cell of Π that contains $\Phi(x, k)$. The set of all pairs of the form $\{z, \Phi(z)\}$, where $z \in P$, is then a face f_R of I, the face *corresponding* to R. An edge of EI belongs to an oriented boundary of f_R if and only if the corresponding bisquare meets R. The correspondence we have just defined is a bijective one between FI and the set of blue-yellow bigons of K.

If K is a gem encoding an imbedding I, then a drawing that represents I may also be taken to represent K. Under this interpretation, the permutations induced by the drawing act on the red edges of K. Each cell of these permutations is the set of red edges of a red-yellow bigon of K. For any red-yellow bigon B with more than two red edges, there are just two cyclic permutations of $B \cap F$ that are restrictions of the permutations induced by a given drawing that represents K, and they are inverses. They share the property of mapping each red edge e in B to an edge adjacent to a yellow neighbour of e. This requirement suffices to determine the pair of them.

2.6 Premaps

Another combinatorial definition of a map has been given by Tutte (1984). His maps are defined in terms of objects called premaps, which correspond to our maps. In this section, we describe Tutte's approach and relate it to gems.

Let S be a set such that $|S|$ is divisible by 4. Let id denote the identity permutation on S, and let θ and φ be permutations on S satisfying the conditions $\theta^2 = \varphi^2 = \mathrm{id}$ and $\theta\varphi = \varphi\theta$, and suppose that $x, \theta x, \varphi x, \theta\varphi x$ are distinct for each $x \in S$. Let P be another permutation on S such that $P\theta = \theta P^{-1}$, and for each x let the cells of P containing x and θx be distinct. Then (S, θ, φ, P) defines a *premap*, M. Tutte also defines Ψ_K as the group of permutations generated by a non-empty set K of permutations of S. He calls the premap M a *map* if for each $x \in S$ and $y \in S$ there is a permutation $\pi \in \Psi_{\{\theta,\varphi,P\}}$ such that $\pi x = y$.

We now show how to construct a gem $K(M)$ that represents a premap $M = (S, \theta, \varphi, P)$, where $S \neq \emptyset$. The vertices of $K(M)$ are the elements of S. For each $x \in S$, let us draw a red edge joining x to θx, a blue edge joining x to φx, and a yellow edge joining x to $P\theta x$. (See Figure 2.13.) Since $\theta^2 = \varphi^2 = \mathrm{id}$ and $(P\theta)^2 = \theta P^{-1} P\theta = \theta^2 = \mathrm{id}$, every vertex is incident with only one edge of a given colour. Hence, the resulting graph is cubic. Moreover, each red-blue bigon is a square, since $\theta\varphi\theta\varphi = \theta^2\varphi^2 = \mathrm{id}$, and so the construction has yielded a gem. Note also that the two cells of P containing x and θx, respectively, correspond in an obvious way to the two orbits of the red-yellow bigon containing the red edge that joins x and θx.

Conversely, for any gem K it is easy to construct a premap M for which $K = K(M)$. S is the vertex set of K, and the images of the red and blue edges under ψK are the cells of the permutations θ and φ, respectively. Moreover, for any vertex x, Px is the unique vertex joined to θx by a yellow edge.

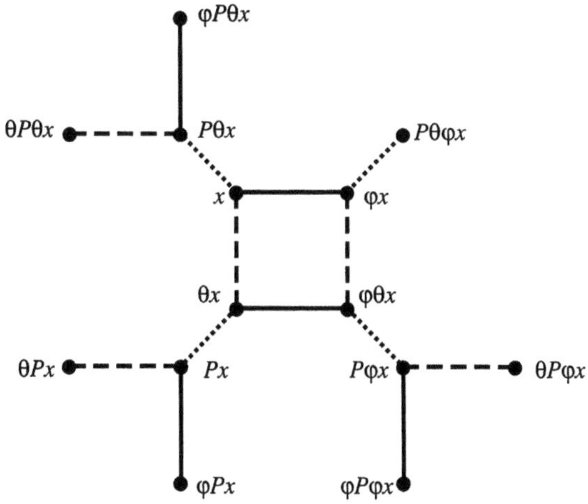

FIGURE 2.13.

Theorem 2.3 *A premap M is a map if and only if $K(M)$ is connected.*

Proof. Suppose that $K(M)$ is connected, and choose $x, y \in VK(M)$. By Theorem 1.4, $K(M)$ has a path Q joining x and y. Each edge of Q is red, blue, or yellow. It follows that for each edge $e \in Q$ there is a permutation in $\{\theta, \varphi, P\theta\}$ that interchanges the ends of e. Hence $y = \pi x$ for some product π of such permutations. Since $\pi \in \Psi_{\{\theta,\varphi,P\}}$, it follows that M is a map.

Conversely, suppose that $K(M)$ is not connected. Choose x and y to be vertices of distinct components of $K(M)$. For any vertex v, θv and φv are adjacent to v and Pv is adjacent to θv. Therefore, x and σx are vertices of the same component of $K(M)$ for any permutation $\sigma \in \Psi_{\{\theta,\varphi,P\}}$. Hence, no permutation of $\Psi_{\{\theta,\varphi,P\}}$ maps x to y. ■

Tutte defines an *edge* of a premap $M = (S, \theta, \varphi, P)$ to be a set of the form $\{x, \theta x, \varphi x, \theta\varphi x\}$, where $x \in S$. The members of such a set are the vertices of a red-blue bigon of $K(M)$. Thus there is a bijection between the set of edges of M and the set of red-blue bigons of $K(M)$.

For each $x \in S$, the pair of cells of P that meet $\{x, \theta x\}$ is called by Tutte a *vertex* of the premap M. In $K(M)$ we recall that x is joined to θx by a red edge and θx to Px by a yellow one. The cell of P that contains x is therefore the set of vertices at an even distance from x in the graph induced by the red-yellow bigon B passing through x. The remaining vertices of VB therefore constitute the cell of P that contains θx. Hence VB is the union of the elements of a vertex of M. Therefore, there is a bijection between the set of vertices of M and the set of red-yellow bigons of $K(M)$.

Next, recall that x is joined to φx by a blue edge, and φx to $P\theta\varphi x$ by a yellow one. Let $P^* = P\theta\varphi$. Then the cell of P^* that contains x is the set of vertices at an even distance from x in the graph induced by the blue-yellow bigon R passing through x. This cell is therefore distinct from the cell of P^* containing φx. Moreover, $\varphi(P^*)^{-1} = \varphi(P\theta\varphi)^{-1} = \varphi^2\theta P^{-1} = P\theta = P\theta\varphi^2 = P^*\varphi$. We conclude that $(S, \varphi, \theta, P^*)$ is a premap, M^*. Tutte calls M^* the premap *dual* to M. Since $P^*\varphi = P\theta$, we observe that the gem $K(M^*)$ is obtained from $K(M)$ by interchanging the colours red and blue. Tutte defines a *face* of M to be a vertex of M^*. Therefore, there is a bijection between the set of faces of M and the set of blue-yellow bigons of $K(M)$.

Now suppose that M is a map. Let us describe two elements of S as *equivalent* if some permutation in $\Psi_{\{\theta\varphi,P\}}$ maps one to the other. Thus x and Px are equivalent for each $x \in S$. If B is the red-yellow bigon in $K(M)$ passing through x, then it follows inductively that at least half of the vertices of B belong to the equivalence class $[x]$ of x. In fact, each red edge of B is incident on a member of $[x]$. Furthermore, $\varphi\theta x \in [x]$. Hence $[x]$ contains at least half of the vertices of the red-yellow bigon passing through $\varphi\theta x$, and each red edge of this bigon is incident on a member of $[x]$. Therefore, the same can be said of any red-yellow bigon and of any red edge, since $K(M)$ is connected by Theorem 2.3. Therefore, $[x]$ contains at least half of the vertices of $K(M)$. We infer that, at most, two equivalence classes exist. Tutte defines M to be *orientable* if two equivalence classes do in fact exist.

Theorem 2.4 *A map M is orientable if and only if $K(M)$ is bipartite.*

Proof. Suppose first that M is an orientable map (S, θ, φ, P). Choose $x \in S$, with equivalence class $[x]$. If Y is the red-blue bigon of $K(M)$ passing through x, then $[x] \cap VY = \{x, \theta\varphi x\}$. Since $[x] \cap VZ \neq \emptyset$ for each red-blue bigon Z of $K(M)$, it follows that every red or blue edge of $K(M)$ has just one end in $[x]$. The same is true for every yellow edge, since $Px \in [x]$ but $\theta x \notin [x]$. We conclude that $K(M)$ is bipartite.

On the other hand, suppose that M is not orientable. Then there is a permutation $\sigma \in \Psi_{\{\theta\varphi, P\}}$ such that $\sigma x = \theta x$ for some $x \in S$. Note that any vertex $y \in S$ is joined to Py and to $\theta\varphi y$ by paths of length 2. Therefore, there is a path of even length joining x to σx. Since x is adjacent to $\theta x = \sigma x$, we find that $K(M)$ contains a circuit of odd length and therefore is not bipartite. ∎

The approach of Tutte has therefore been reconciled with ours, except that his premaps are our maps. His premaps can therefore also be represented by drawings in which points represent vertices and curves represent edges in the same manner as before. Since each red edge of a gem is specified by its pair of ends, we may take the permutations induced by a drawing representing a premap (S, θ, φ, P) to be permutations of the set of all pairs of the form $\{x, \theta x\}$, where $x \in S$. One such permutation maps $\{x, \theta x\}$ to $\{Px, \theta Px\}$ and, since $\theta P \theta x = \theta^2 P^{-1} x = P^{-1} x$ and $P\theta x = \theta P^{-1} x$, the other maps $\{x, \theta x\}$ to $\{P^{-1}x, \theta P^{-1}x\}$. These two permutations are therefore inverses of each other.

3

Classification of Surfaces

Tutte (1984) approached the classification of surfaces from a combinatorial viewpoint. In this chapter, we show how the classification of surfaces by means of 3-graphs follows from Tutte's approach and the relationship between 3-graphs and premaps.

In view of Theorem 2.4, a 3-graph is said to be *orientable* if it is bipartite, and *non-orientable* otherwise. The *orientability* of a 3-graph K is defined to be 0 if K is orientable, and 1 otherwise.

We let $\mathcal{B}(K)$, $\mathcal{Y}(K)$, and $\mathcal{R}(K)$ denote the sets of red-yellow, red-blue, and blue-yellow bigons, respectively, in a 3-graph K. Let $r(K)$ be the total number of bigons in K. We define the *Euler characteristic* of K to be

$$\chi(K) = r(K) - \frac{|VK|}{2},$$

which gives $\chi(K) = |\mathcal{B}(K)| - |\mathcal{Y}(K)| + |\mathcal{R}(K)|$ when K is a gem since $|VK| = 4|\mathcal{Y}(K)|$. The *Euler characteristic* of a map M is $\chi(K(M))$.

Tutte defines a surface as the class of all maps with a given Euler characteristic and given orientability, provided that the class is non-empty. In our setting, such a class corresponds to a class of connected gems. However, we will work in the more general setting of 3-graphs and define a *surface* as the class of all connected 3-graphs with a given Euler characteristic and a given orientability, provided that the class is non-empty. The main theorem of this chapter classifies all such surfaces. It states that one 3-graph can be obtained from another by a finite number of "moves" if and only if they belong to the same surface, where the moves are the crystallisation moves of Ferri and Gagliardi (1982) and are defined in the next section.

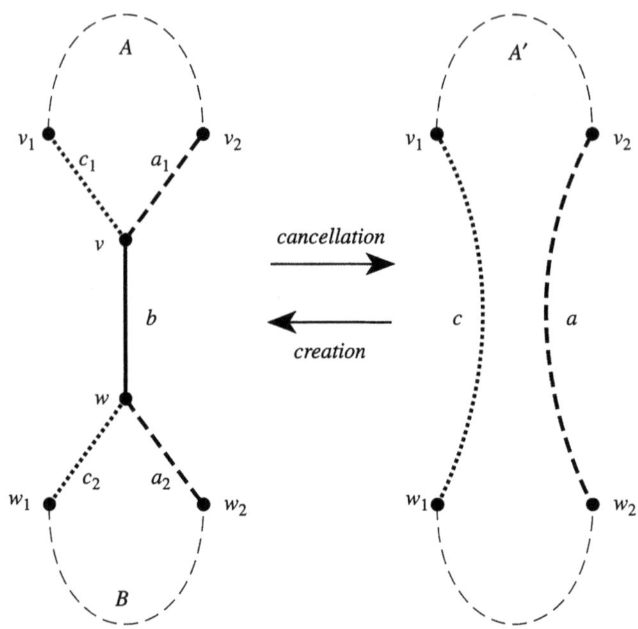

FIGURE 3.1.

3.1 Dipoles

Let v and w be a pair of adjacent vertices in a 3-graph K. Suppose that v and w are joined by one edge b, which is blue. Following Ferri and Gagliardi (1982), we say that b is a *blue* 1-*dipole* if the red-yellow bigons A and B passing through v and w, respectively, are distinct. Let c_1 and c_2 be the yellow edges incident on v and w, respectively, and let a_1 and a_2 be the red edges incident on v and w, respectively. Let v_1, v_2, w_1, w_2 be the vertices other than v and w incident on c_1, a_1, c_2, a_2, respectively. (See Figure 3.1.) The *cancellation* of this blue 1-dipole b is the operation of deletion of the vertices v and w followed by the insertion of a yellow edge c joining v_1 to w_1 and a red edge a joining v_2 to w_2. We denote the resulting 3-graph by $K - [b]$. We observe that A and B have coalesced into one red-yellow bigon A'. The *creation* of a blue 1-dipole is the inverse operation. Similar definitions can be made for *red* and *yellow* 1-*dipoles* and their cancellations and creations.

Now suppose that v and w are joined by two edges a and b coloured red and blue, respectively. Following Ferri and Gagliardi (1982), we say that $\{a, b\}$ is a *red-blue* 2-*dipole* if the yellow edges c_1 and c_2 incident on v and w, respectively, are distinct. Let c_1 join v to v_1 and let c_2 join w to w_1. (See Figure 3.2.) The *cancellation* of this red-blue 2-dipole is the operation of deletion of the vertices v and w followed by the insertion of a yellow edge c joining v_1 to w_1. We denote the resulting 3-graph by $K - [a, b]$. We observe that c_1 and c_2 have coalesced into one yellow edge c. The *creation* of a red-blue 2-dipole is the inverse operation.

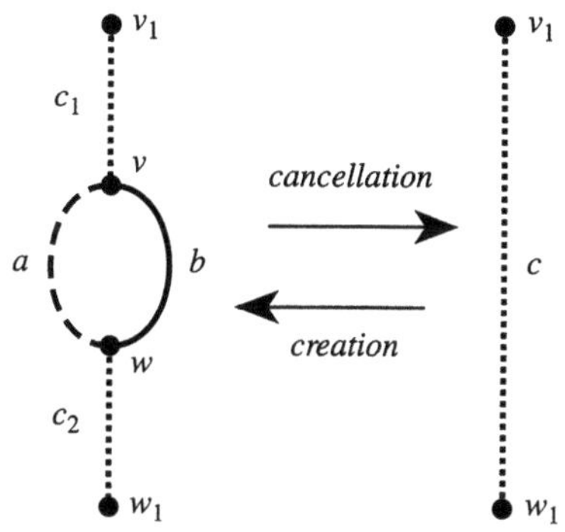

FIGURE 3.2.

Similar definitions can be made for *red-yellow* and *blue-yellow 2-dipoles* and their cancellations and creations.

We note that the yellow edge c_1 is a yellow 1-dipole in K and that the 3-graph $K - [c_1]$ is isomorphic to $K - [a, b]$. Hence, cancellation or creation of a 2-dipole is in fact a special case of a 1-dipole cancellation or creation.

A μ*-move* is a cancellation or creation of a 1-dipole. Two 3-graphs are μ*-equivalent* if one can be obtained from the other by a finite sequence of μ-moves. It is shown in Ferri and Gagliardi (1982) that two 3-graphs are equivalent if and only if the corresponding surfaces are homeomorphic. Thus, the following theorem is equivalent to the classification of surfaces, due to Dehn and Heegaard (1907). Our proof is essentially theirs translated into the setting of coloured graphs. A similar proof in terms of premaps appears in Tutte (1984). The proof of the necessity appears as Lemma 3.2, and the proof of the sufficiency appears in Sections 3.2 and 3.3.

Theorem 3.1 *Two connected 3-graphs are μ-equivalent if and only if they have the same Euler characteristic and orientability.*

Lemma 3.2 *Connected μ-equivalent 3-graphs belong to the same surface.*

Proof. Let K and L be μ-equivalent connected 3-graphs. We may assume that K is obtained from L by cancellation of a 1-dipole. We will show that $\chi(L) = \chi(K)$, and that L is bipartite if and only if K is bipartite. Indeed, in a 1-dipole cancellation, the number of bigons drops by one and the number of vertices by two. Therefore

$$\begin{aligned} \chi(K) &= r(K) - \frac{|VK|}{2} \\ &= r(L) - 1 - \frac{|VL|}{2} + 1 \\ &= \chi(L). \end{aligned}$$

Now, assuming that L is bipartite, one may colour the vertices of L black or white so that adjacent vertices receive distinct colours. Using the notation of Figure 3.1,

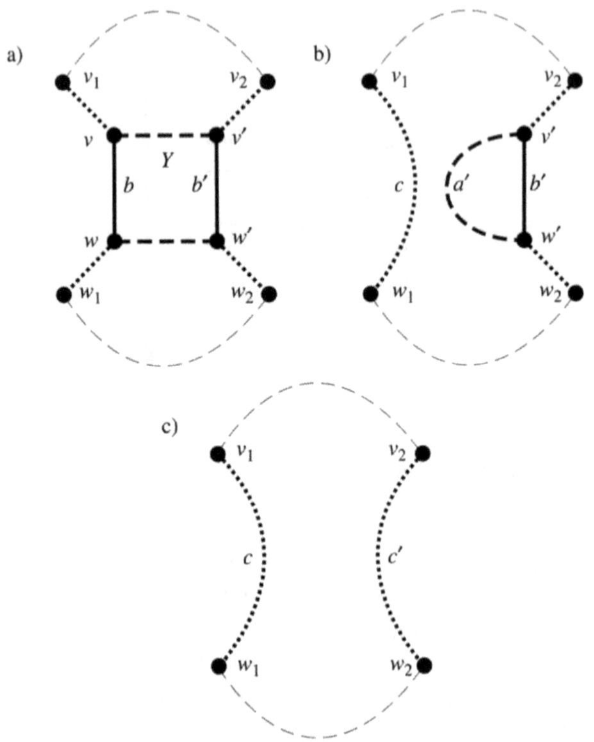

FIGURE 3.3.

evidently v and w receive distinct colours, as do v_1 and w_1, and v_2 and w_2. Hence, we conclude that K is bipartite. Similarly, L is bipartite if K is bipartite. ■

We conclude this section by describing another operation on 3-graphs called "cancellation" of a red-blue bisquare. This operation is in fact a pair of dipole cancellations.

Suppose that Y is a red-blue bisquare in a 3-graph K. The following uses the notation of Figure 3.3a. If b is a blue 1-dipole, then let $K' = K - [b]$. (See Figure 3.3b.) Let a' denote the red edge of K' that joins v' to w'. If $\{a', b'\}$ is a red-blue 2-dipole, then let $K'' = K' - [a', b']$. (See Figure 3.3c.) We say that K'' is obtained from K by *cancellation* of the red-blue bisquare Y. Let c and c' denote the yellow edges that join v_1 to w_1, and v_2 to w_2, respectively. Note that they belong to a common red-yellow bigon. The inverse operation is described as *splitting* c and c' to create the red-blue bigon Y. By definition, K and K'' are μ-equivalent.

3.2 Reduced and unitary 3-graphs

Suppose that K is a connected 3-graph with at least two red-yellow bigons. Since K is connected, there must exist a blue 1-dipole in K. Cancelling this dipole reduces the number of red-yellow bigons by one. Proceeding inductively, we obtain a connected 3-graph that has exactly one red-yellow bigon. Similarly, we reduce to one the numbers of blue-yellow and red-blue bigons by red 1-dipole and yellow 1-dipole cancellations. The resulting 3-graph is a *reduced* 3-graph of K, and has just three bigons, one of each type. Note that a reduced 3-graph of K is μ-equivalent to K.

The 3-graph with just two vertices is *trivial.*

Lemma 3.3 *If K is a connected 3-graph, then $\chi(K) \leq 2$. Moreover, if $\chi(K) = 2$, then K is bipartite and the only reduced 3-graph of K is trivial.*

Proof. Let K' denote a reduced 3-graph of K. Since K and K' are μ-equivalent, we have $\chi(K) = \chi(K')$. Furthermore,

$$\begin{aligned}\chi(K') &= r(K') - \frac{|VK'|}{2} \\ &\leq 2\end{aligned}$$

since all 3-graphs have at least two vertices. If $\chi(K) = \chi(K') = 2$, then $|VK'| = 2$, and K' is the trivial 3-graph. Since the trivial 3-graph is bipartite, we have that K is bipartite by Lemma 3.2. ■

The *(combinatorial) sphere* is the class of all connected 3-graphs with Euler characteristic 2, and all 3-graphs in it are called *spherical.* Thus, the trivial 3-graph is spherical. An arbitrary 3-graph is said to be *spherical* if each component is spherical. Thus, a 3-graph K is spherical if and only if $\chi(K) = 2c(K)$.

A connected 3-graph K is *unitary* if $|\mathcal{B}(K)| = |\mathcal{R}(K)| = 1$. Hence, reduced 3-graphs are unitary.

Lemma 3.4 *A connected 3-graph is μ-equivalent to the trivial 3-graph or a unitary gem.*

Proof. Let K' denote a reduced 3-graph of a connected 3-graph K. If K is spherical, then K' is the trivial 3-graph by Lemma 3.3, as required. Hence, we assume otherwise. Let $\mathcal{B}(K') = \{B\}$ and $\mathcal{Y}(K') = \{Y\}$. If Y is a digon, then each yellow edge incident on a vertex of Y is a yellow 1-dipole, a contradiction. If Y is a bisquare, then K' is a unitary gem and we are done. In the remaining case, let a_1 and a_2 be red edges of Y both adjacent to a common blue edge b. Let P_1 and P_2 be the two paths of $B - \{a_1, a_2\}$. The two possibilities are illustrated in Figure 3.4a. Clearly, there exist yellow edges $c_1 \in P_1$ and $c_2 \in P_2$, and therefore we split c_1 and c_2 to create a red-blue bisquare Y_1. Let K'' denote the resulting graph. (See Figure 3.4b.) Evidently, b is a blue 1-dipole in K'', and therefore we cancel it to obtain a unitary 3-graph U. (See Figure 3.4c.) The red-blue bigon in U corresponding to Y has one blue edge less than Y. Proceeding inductively, we obtain a unitary 3-graph U', where the red-blue bigon corresponding to Y is a bisquare. However, U' is a unitary gem since all new red-blue bigons created were bisquares. ■

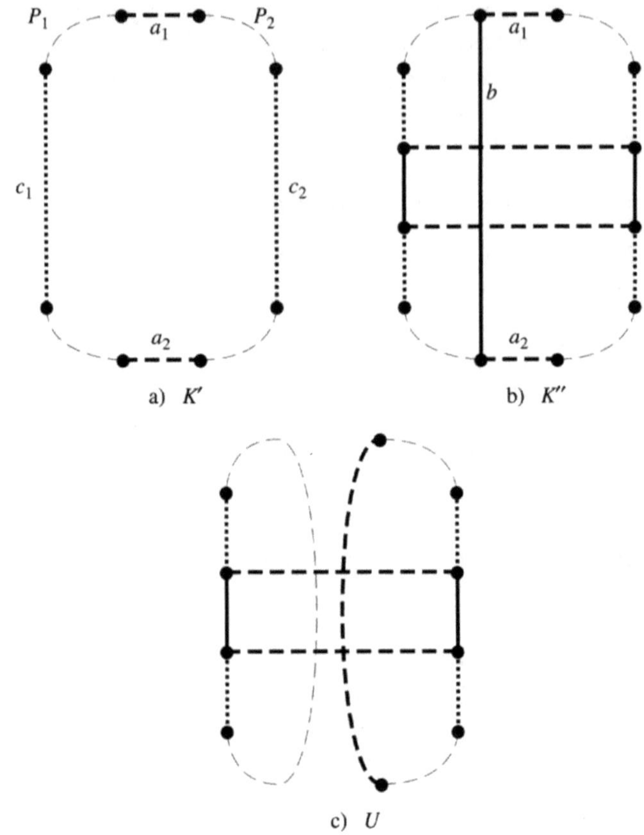

FIGURE 3.4.

We now investigate the case where the surface is not the combinatorial sphere. By means of μ-moves, we transform an arbitrary 3-graph in such a surface into a unitary gem U.

Let B be the one red-yellow bigon in U and let R be the set of red edges in B. Let Y be an arbitrary red-blue bigon of U. If $B \cup Y$ includes a circuit C such that $|C \cap Y \cap R| = 1$, then we say that Y is a *cross-cap* of B. (See Figure 3.5a.) It is an *assembled* cross-cap of B if there exists a subpath of B with just two red edges, both in Y. (See Figure 3.5b.) If Y is not a cross-cap, then it is a *cap* of B.

Suppose that there exist two caps X and Y of B. If $B \cup X \cup Y$ includes a circuit C such that $|C \cap Y \cap R| = 1$, then we say that X and Y are *bound* in B, and $\{X, Y\}$ is a *handle* of B. (See Figure 3.6a.) Such a handle is *assembled* if there exists a subpath of B with just four red edges, all in $X \cup Y$. (See Figure 3.6b.)

Lemma 3.5 *A given unitary gem is μ-equivalent to another for which all of the cross-caps are assembled.*

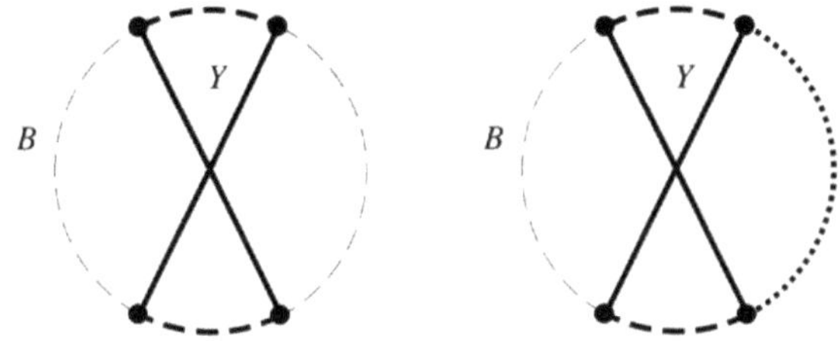

a) b)

FIGURE 3.5.

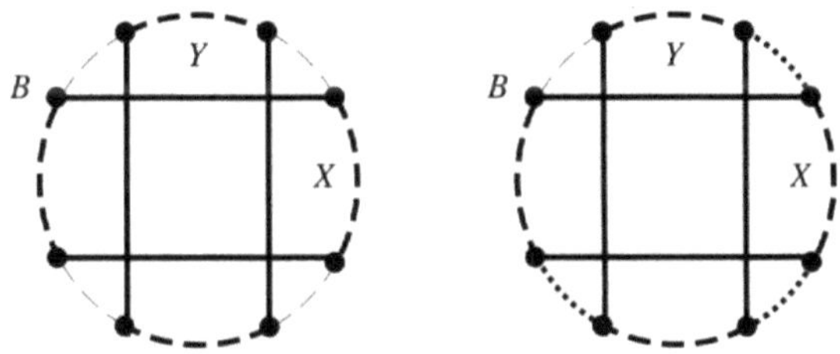

a) b)

FIGURE 3.6.

Proof. Let B be the one red-yellow bigon in a unitary gem U and let Y be an unassembled cross-cap. Label the edges of B incident on vertices of VY as in Figure 3.7a. Split d and d' to create a red-blue bigon X and let U' denote the resulting gem. (See Figure 3.7b.) In U', a and a' belong to distinct red-yellow bigons. Hence, we may let U'' denote the gem obtained by cancelling Y in U'. (See Figure 3.7c.) Clearly, $|\mathcal{B}(U'')| = |\mathcal{B}(U)| = 1$ and $|\mathcal{R}(U'')| = |\mathcal{R}(U)| = 1$. It is also clear that X is an assembled cross-cap in U''. Moreover, no assembled cross-cap in U has been lost; the red edges of such a cross-cap are in $B - \{c, a, d, c', a', d'\}$.

By this procedure, we can assemble every cross-cap. ■

Lemma 3.6 *Let U be a unitary gem with all cross-caps assembled. Then all caps in U are bound.*

Proof. Suppose that there exists an unbound cap Y in U. Label the vertices and edges adjacent to Y as in Figure 3.8. Since no blue edge in U joins a vertex in $VB_v[v', w']$ to a vertex in $VB_{v'}[v, w]$, b and b' must belong to distinct blue-yellow bigons, contradicting the fact that U is unitary. ■

In a unitary gem with all cross-caps assembled, either the red-blue bigons are all assembled cross-caps or there is a handle. We now assemble handles.

Lemma 3.7 *Any unitary gem is μ-equivalent to one in which each red-blue bigon is an assembled cross-cap or a member of an assembled handle.*

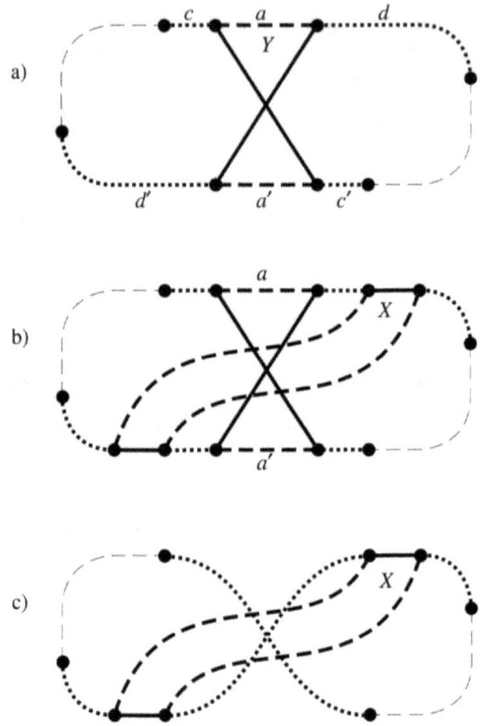

FIGURE 3.7.

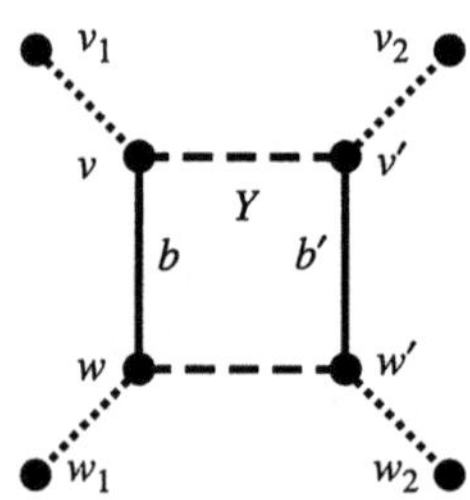

FIGURE 3.8.

Proof. Suppose that there exists a handle $\{X, Y\}$ in a unitary gem U. The following uses the notation in Figure 3.9a. Split d_2 and d_4 to create a red-blue bigon W and let U_1 denote the resulting gem. (See Figure 3.9b.) In U_1, a_1 and a_3 are in distinct red-yellow bigons, and so X may be cancelled to yield a gem U_2. (See Figure 3.9c.) In U_2, let c_5 denote the yellow edge other than c_4 adjacent to a_4. Let b_5 denote the blue edge of W adjacent to c_5, and let d_5 denote the yellow edge other than c_5 adjacent to b_5. Split c_5 and d_5 to create a red-blue bigon Z and let U_3 denote the resulting gem. (See Figure 3.9d.) In U_3, a_2 and a_4 are in distinct red-yellow bigons, and so we may cancel Y to obtain a gem U'. (See Figure 3.9e.)

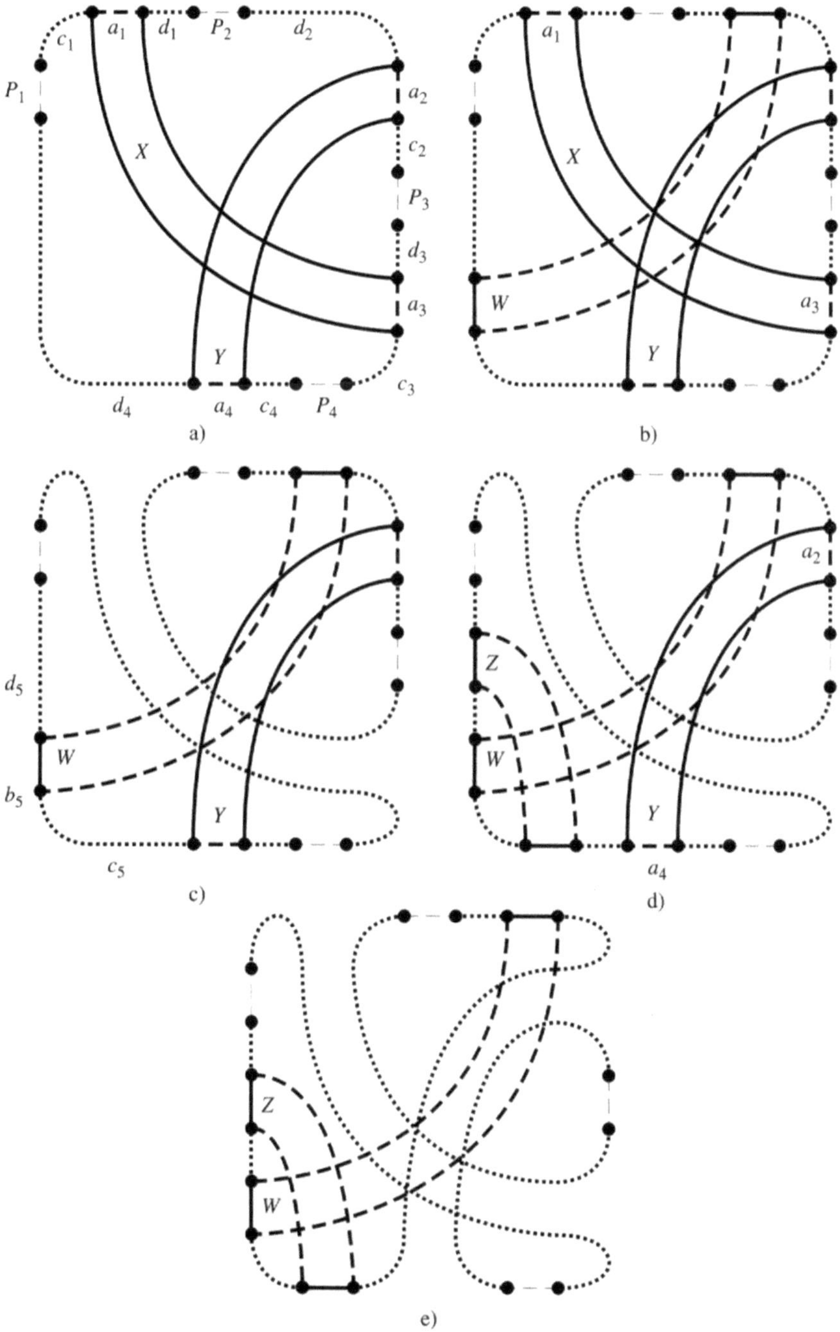
c_1
a_1
d_1
P_2
d_2
P_1
a_2
c_2
P_3
d_3
a_3
X
Y
c_3
d_4
a_4
c_4
P_4
a)
a_1
X
W
a_3
Y
b)
d_5
W
b_5
Y
c_5
c)
a_2
Z
W
Y
a_4
d)
Z
W
e)

FIGURE 3.9.

The above process transforms U into another unitary gem U'. The handle $\{X, Y\}$ has been replaced by the assembled handle $\{W, Z\}$. The red edges in any assembled cross-cap or in a member of any other assembled handle are in

$$B - \{a_1, c_1, d_1, a_2, c_2, d_2, a_3, c_3, d_3, a_4, c_4, d_4\}.$$

These cross-caps and handles therefore remain assembled.

By repeating this process, we assemble all of the handles and produce the required unitary gem. No red-blue bigons remain, by Lemma 3.6. ■

Lemma 3.8 *Any unitary gem is μ-equivalent to one in which each red-blue bigon is an assembled cross-cap or each red-blue bigon is a member of an assembled handle.*

Proof. We may suppose that our unitary gem U is already in the form specified in Lemma 3.7. Suppose it to have at least one assembled cross-cap and at least one assembled handle. Then the situation arising is depicted in Figure 3.10a, which displays a handle $\{X, Y\}$ and a cross-cap Z containing a red edge a_2 that has a yellow neighbour in common with a red edge a_3 of Y. Split c and d to create a red-yellow bigon W and let U_1 denote the resulting graph. (See Figure 3.10b.) In U_1, a_1 and a_2 are in distinct red-yellow bigons, and so we may cancel Z to obtain a gem U_2. (See Figure 3.10c.) In U_2, split c' and d' to create a red-blue bigon V, and let U_3 denote the resulting gem. (See Figure 3.10d.) In U_3, a_4 and a_6 are in distinct red-yellow bigons, and so X may be cancelled to yield a unitary gem U_4. (See Figure 3.10e.)

In U_4, we have an assembled cross-cap V, an unassembled cross-cap Y, and a cap W. The original assembled handles and cross-caps remain assembled.

Our next step is to replace the cross-cap Y by an assembled cross-cap, as in Lemma 3.5. The assembled cross-caps and handles of U_4 are preserved. By Lemma 3.6, the red-blue bigon W is transformed into another cross-cap, which we assemble. We thus obtain a unitary gem of the form in Lemma 3.7, but with one less handle and two more cross-caps. Repetition of the above procedure gives a unitary gem in which all red-blue bigons are cross-caps. ■

3.3 Canonical gems

A 3-graph is *canonical* if it is trivial or a unitary gem U in which $\mathcal{Y}(U)$ consists entirely of assembled cross-caps or entirely of members of assembled handles.

The trivial 3-graph is orientable, as is a unitary canonical 3-graph whose red-blue bigons are members of handles. The *genus* of such a 3-graph is the number of its handles. Thus, the genus of the trivial 3-graph is 0. On the other hand, a unitary canonical 3-graph whose red-blue bigons are cross-caps is non-orientable. The *cross-cap number* of such a 3-graph is the number of its cross-caps.

Proof of Theorem 3.1. That connected μ-equivalent 3-graphs have the same Euler characteristic and orientability follows from Lemma 3.2. Let K and L therefore

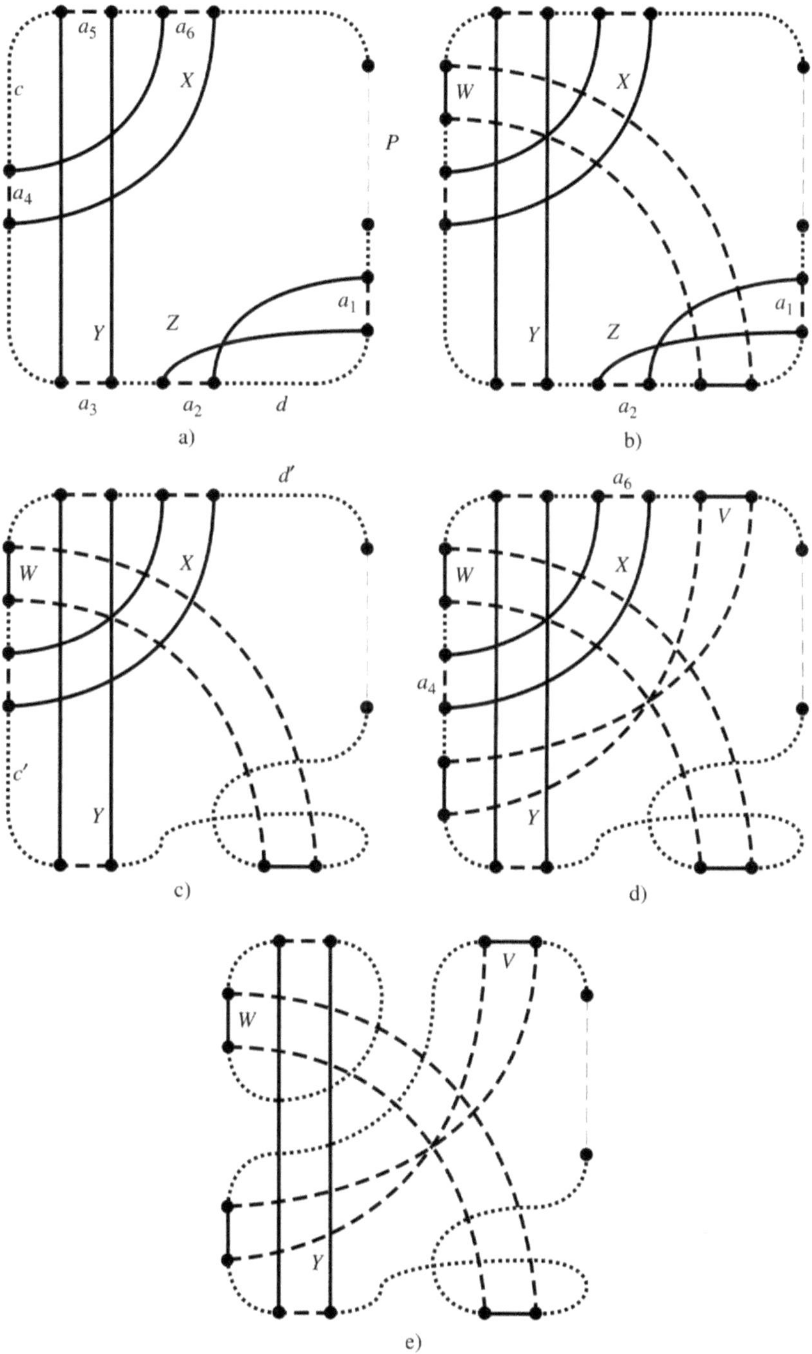

FIGURE 3.10.

be connected 3-graphs with the same Euler characteristic and the same orientability. By Lemmas 3.4 and 3.8, we find that K and L are μ-equivalent to canonical 3-graphs K' and L', respectively.

Suppose first that K and L are non-orientable. Then each red-blue bigon of K' or L' is a cross-cap. Let k be the number of cross-caps in K'. Then K' has $k+2$ bigons and $4k$ vertices, so that

$$\begin{aligned}\chi(K') &= k+2-2k \\ &= 2-k.\end{aligned}$$

Similarly, $\chi(L') = 2 - j$, where j is the number of cross-caps in L'. Thus, $2-j = \chi(L') = \chi(L) = \chi(K) = \chi(K') = 2-k$, and so we have $k = j$. Hence K' and L' are isomorphic, and K is μ-equivalent to L.

If K and L are orientable, then no red-blue bigon of K' or L' is a cross-cap. Let g be the number of handles in K'. If $g = 0$, then K' is trivial, so that $\chi(L') = \chi(K') = 2$. By Lemma 3.3, it follows that L' is also trivial, and we infer that K is μ-equivalent to L. Suppose therefore that $g > 0$. Then K' has $2g+2$ bigons and $8g$ vertices. Hence

$$\begin{aligned}\chi(K') &= 2g+2-4g \\ &= 2-2g,\end{aligned}$$

and the argument is completed as in the non-orientable case. ■

The genus g of an orientable canonical gem is also called the *genus* of its surface and of any other 3-graph in that surface. The proof above shows that if K is such a 3-graph, then $\chi(K) = 2-2g$. Likewise, the cross-cap number k of an non-orientable canonical gem is also called the *cross-cap number* or *genus* of its surface and of any other 3-graph in that surface. If L is such a 3-graph, then $\chi(L) = 2-k$. A surface is called *orientable* or *non-orientable surface* depending on whether the 3-graphs in it are orientable or non-orientable. We can now say that there is just one orientable surface whose genus is a given non-negative integer, and just one non-orientable surface whose cross-cap number is a given positive integer and, moreover, there are no other surfaces.

We would like to mention that Vince (private communication) has a similar method for obtaining the classification of surfaces. It is of interest to note that his canonical 3-graphs are different from ours.

3.4 Planar graphs

We say that a graph is *imbeddable* in a surface if it is the graph of a gem in that surface. It is of some interest to find the surface of minimum genus in which a given graph is imbeddable. Accordingly, we define the *genus* of a non-empty connected graph G to be the minimum genus of a gem underlaid by G. The *genus* of the empty connected graph is defined to be 0. A connected graph of genus 0 is said to be *planar*. In general, a graph is *planar* if all of its components are planar. Thus, non-empty connected planar graphs are the graphs imbeddable in the sphere. They are the ones that underlie spherical gems.

In general, the determination of the genus of a graph is an unsolved problem. However, a number of characterisations of planar graphs are known, and we describe several in this book. The reader is referred to Gross and Tucker (1987) for an account of techniques used to find the genus of a graph.

The following theorem is immediate from Lemma 3.3 and the fact that $|VK| = 4|\mathcal{Y}(K)|$ for any gem K.

Theorem 3.9 *Let K be a connected gem. Then*

$$|\mathcal{R}(K)| \leq |\mathcal{Y}(K)| - |\mathcal{B}(K)| + 2,$$

and equality holds if and only if K is spherical.

The *genus* or *cross-cap number* of an imbedding is that of the gem that encodes it. An imbedding of genus 0 is *planar*.

By translating Theorem 3.9 into terms of graphs, we obtain the following corollary.

Corollary 3.10 *Let I be an imbedding of a non-empty connected graph. Then*

$$|FI| \leq |EI| - |VI| + 2,$$

and equality holds if and only if I is planar.

At this stage, it may seem to the reader that our concepts of surface, orientability, and Euler characteristic are rather far removed from the traditional ones. A reconciliation is the objective of the next two chapters.

4

Consistent and Coherent Orientations

In Chapter 3 we classified surfaces according to their Euler characteristic and orientability. In Chapter 5 we will explain the significance of the Euler characteristic. In the present chapter we do the same for orientability, and we also study further properties of this concept. Some of the ideas developed here resurface in Chapter 9.

4.1 Orientations

Let G be a graph, and let $C \subseteq EG$. An *orientation* of C is a mapping ρ_C from C into VG such that e is incident on $\rho_C(e)$ for each $e \in C$. Thus, if ρ_C and σ_C are orientations of C and e is a loop in C, then $\rho_C(e) = \sigma_C(e)$. For any orientation ρ_C of C, we define its *conjugate* as the orientation ρ_C^* such that $\rho_C^*(e) \neq \rho_C(e)$ for each link $e \in C$. Thus $\rho_C^{**} = \rho_C$.

If $C = EG$, then an orientation ρ_C of C is also called an *orientation* of G. In this case we refer to $\rho_C(e)$ as the *sense* of an edge $e \in EG$ under ρ_C. We say that $\rho_C^*(e)$ is obtained from $\rho_C(e)$ by *reversing* the sense of e. If e joins vertices u and v, where $\rho_C(e) = v$, then we say that e is *directed from u to v* (under ρ_C). If S and T are disjoint subsets of VG such that $u \in S$ and $v \in T$, then we also say that e is *directed from S to T*. The number of edges directed to a vertex w is called the *indegree* of w in G under ρ_C. The *outdegree* of w in G under ρ_C is the number of edges directed from w.

We describe a vertex v of G as a *source* (with respect to C and ρ_C) if v is not the image of an edge of C under ρ_C, and a *sink* if $v = \rho_C(e)$ for each edge e of C incident on v.

Let $\mathcal{C}$ be a collection $\{C_1, \ldots, C_k\}$ of subsets of EG. An *orientation* ρ of $\mathcal{C}$ is a set $\{\rho_1, \ldots, \rho_k\}$ such that ρ_i is an orientation of C_i for each i. We also define $\rho^* = \{\rho_1^*, \ldots, \rho_k^*\}$.

Let σ be another orientation $\{\sigma_1, \ldots, \sigma_k\}$ of $\mathcal{C}$, where σ_i is an orientation of C_i for each i. We say that ρ and σ are *compatible* (with each other) if $\sigma_i \in \{\rho_i, \rho_i^*\}$ for each i. The relation of being compatible is clearly symmetric.

Example 4.1 *Let G be the graph displayed in Figure 4.1a. For each $i \in J_3$, let $C_i = \{e_i, e_{i+1}\}$. Let ρ_i be the orientation of C_i such that $\rho_i(e_i) = u_i$ and $\rho_i(e_{i+1}) = v_{i+1}$. Then $\{\rho_1, \rho_2, \rho_3\}$ is an orientation of the set $\mathcal{C} = \{C_1, C_2, C_3\}$.*

Now let σ be the orientation $\{\sigma_1, \sigma_2, \sigma_3\}$ of $\mathcal{C}$, where σ_i is an orientation of C_i for each $i \in J_3$, such that $\sigma_1(e_1) = v_1$, $\sigma_1(e_2) = \sigma_2(e_2) = u_2$, $\sigma_2(e_3) = \sigma_3(e_3) = v_3$, and $\sigma_3(e_4) = u_4$. (See Figure 4.1b.) Then $\sigma_1 = \rho_1^*$, $\sigma_2 = \rho_2$, and $\sigma_3 = \rho_3^*$. Hence ρ and σ are compatible.

Finally, let τ be the orientation $\{\tau_1, \tau_2, \tau_3\}$ of $\mathcal{C}$, where τ_i is an orientation of C_i for each $i \in J_3$, such that $\tau_1(e_1) = u_1$, $\tau_1(e_2) = u_2$, $\tau_2 = \rho_2$, and $\tau_3 = \rho_3$. (See Figure 4.1c.) Then ρ and τ are not compatible since $\tau_1 \neq \rho_1$ and $\tau_1 \neq \rho_1^*$.

We say that $\mathcal{C}$ is *coherently oriented under* ρ if for all distinct i and j we have $\rho_i(e) \neq \rho_j(e)$ for each $e \in C_i \cap C_j$. Such an orientation ρ is a *coherent* orientation of $\mathcal{C}$. Thus $\mathcal{C}$ cannot be coherently oriented under ρ if there is a link contained in more than two elements of $\mathcal{C}$ or a loop contained in more than one. Since a loop has only one sense, we shall confine our attention henceforth to graphs with no loops.

Similarly, we say that $\mathcal{C}$ is *consistently oriented under* ρ if for all i and j we have $\rho_i(e) = \rho_j(e)$ for each $e \in C_i \cap C_j$. Such an orientation ρ is a *consistent*

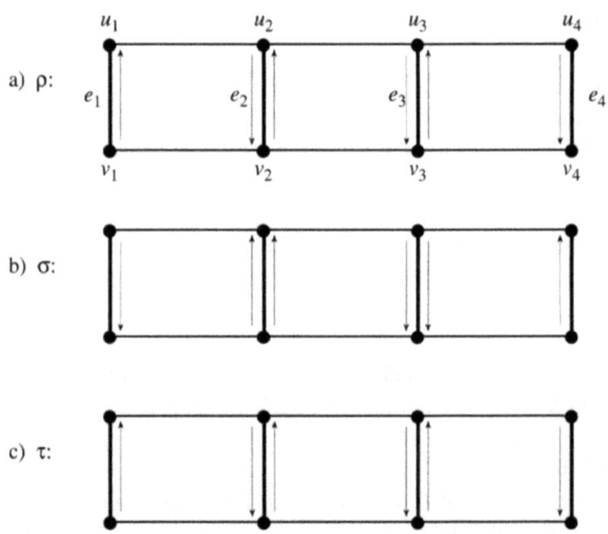

FIGURE 4.1.

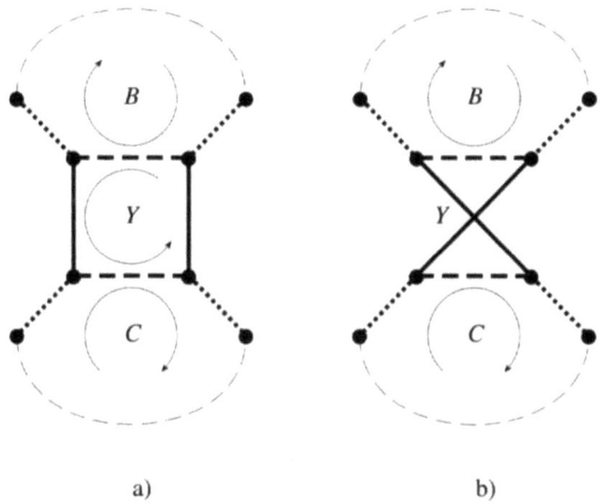

FIGURE 4.2.

orientation of $\mathcal{C}$. By abuse of notation, we write $\rho(e) = \rho_i(e)$ if ρ is a consistent orientation of $\mathcal{C}$. In this way, ρ gives an orientation for $G[\bigcup \mathcal{C}]$.

Example 4.2 *Let K be a gem that encodes an imbedding $I = (X, \pi, \varphi, \iota)$ of a graph G. Choose an edge e in EG joining vertices v and w. Then v and w are represented in K by red-yellow bigons B and C, respectively, and e by a red-blue square Y that meets B and C. Let b and c be the unique edges of $Y \cap B$ and $Y \cap C$, respectively. Recall that each vertex of K is an ordered pair whose second component is 1 or -1. Let ρ_B and ρ_C be bijective orientations of B and C, respectively, such that vertices $\rho_B(b)$ and $\rho_C(c)$ both have a second component 1. If $\iota(e) = 1$, then there exists a bijective orientation ρ_Y of Y such that $\{B, C, Y\}$ is consistently oriented under $\{\rho_B, \rho_C, \rho_Y\}$. (See Figure 4.2a, where the orientations are indicated by the arrows.) If $\iota(e) = -1$, then no such bijective orientation of Y exists. (See Figure 4.2b.)*

Note also that if ρ is a coherent (consistent) orientation of $\mathcal{C}$, then so is ρ^*.

We define the ordered pair $(\mathcal{C}, \rho)$ to be *coherently (consistently) orientable* if there exists a coherent (consistent) orientation of $\mathcal{C}$ compatible with ρ. Under this condition, we may also say that the sets in $\mathcal{C}$ are *coherently (consistently) orientable* with respect to ρ. If $(\mathcal{C}, \rho)$ is coherently (consistently) orientable, then so is $(\mathcal{D}, \tau)$ if $\mathcal{D} \subseteq \mathcal{C}$ and $\tau \subseteq \rho$. In any case, the ordered pair $(\mathcal{C}, \rho)$ is called a *net*.

Observe that if $|\mathcal{C}| = 2$, then coherent and consistent orientability (with respect to ρ) are equivalent. In this case, $\mathcal{C} = \{C_1, C_2\}$ and $\rho = \{\rho_1, \rho_2\}$, where ρ_i is an orientation of C_i for each $i \in J_2$, and $(\mathcal{C}, \rho)$ is coherently orientable if and only if there are no two edges a, b in $C_1 \cap C_2$ such that $\rho_1(a) = \rho_2(a)$ but $\rho_1(b) \neq \rho_2(b)$. Suppose that $(\mathcal{C}, \rho)$ is indeed coherently orientable. Then ρ is compatible with a coherent orientation σ of $\mathcal{C}$. Here $\sigma = \{\sigma_1, \sigma_2\}$, where σ_i is an orientation of C_i for each $i \in J_2$. If $C_1 \cap C_2 \neq \emptyset$, we say that the orientations σ_1 and σ_2 *induce* each other.

4.2 Pairwise coherently orientable nets

Let $\mathcal{C}$ be a collection of sets. We define $H(\mathcal{C})$ to be the simple graph with vertex set $\mathcal{C}$ such that two vertices are adjacent if and only if they meet. If $\mathcal{C}$ is a collection of subsets of the edge set of a graph G, we also say that the net $(\mathcal{C}, \rho)$ is *pairwise coherently orientable* if any two sets in $\mathcal{C}$ are coherently orientable with respect to the orientation ρ of $\mathcal{C}$. Furthermore, we say that $\mathcal{C}$ is *pairwise coherently orientable* if there exists an orientation ρ of $\mathcal{C}$ such that $(\mathcal{C}, \rho)$ is pairwise coherently orientable. These concepts feature in the following criterion for the coherent or consistent orientability of nets.

Theorem 4.1 *Let $(\mathcal{C}, \rho)$ be a pairwise coherently orientable net in a graph G. Suppose further that $(\mathcal{D}, \tau)$ is coherently (consistently) orientable whenever $\mathcal{D}$ is the vertex set of an induced circuit in $H(\mathcal{C})$ and $\tau \subseteq \rho$. Then $(\mathcal{C}, \rho)$ is coherently (consistently) orientable.*

Proof. Clearly, we can assume that $H(\mathcal{C})$ is connected. Let $\mathcal{C} = \{C_1, \ldots, C_p\}$ and $\rho = \{\rho_1, \ldots, \rho_p\}$, where ρ_i is an orientation of C_i for each i. Since $H(\mathcal{C})$ is connected, we may also assume without loss of generality that the subgraph H_j of $H(\mathcal{C})$ spanned by $\{C_1, \ldots, C_j\}$ is connected for each j.

Assume as an inductive hypothesis that, for some k, $\{\sigma_1, \ldots, \sigma_k\}$ is a coherent (consistent) orientation of $\{C_1, \ldots, C_k\}$ compatible with $\{\rho_1, \ldots, \rho_k\}$, where σ_i is an orientation of C_i for each i. (Clearly, we may take $\sigma_1 = \rho_1$ if $k = 1$.) If $k = |\mathcal{C}|$, the theorem is true. If $k \neq |\mathcal{C}|$, there exists a vertex C_{k+1} in $VH(\mathcal{C}) - \{C_1, \ldots, C_k\}$. Because H_{k+1} is connected, C_{k+1} is joined by an edge $e_i \in EH(\mathcal{C})$ to some vertex $C_i \in \{C_1, \ldots, C_k\}$.

Suppose that $(\{C_1, \ldots, C_{k+1}\}, \{\rho_1, \ldots, \rho_{k+1}\})$ is not coherently (consistently) orientable. Since $(\mathcal{C}, \rho)$ is pairwise coherently orientable, it follows that C_{k+1} is joined by an edge $e_j \in EH(\mathcal{C})$ to a second vertex $C_j \in \{C_1, \ldots, C_k\}$ such that the orientations induced on C_{k+1} by σ_i and σ_j differ. Because H_k is connected, it contains a path P, of minimal length, joining C_i to C_j. We may assume that C_i and C_j are chosen so that no other vertex of P is adjacent to C_{k+1} in $H(\mathcal{C})$. Then $P \cup \{e_i, e_j\}$ is an induced circuit in $H(\mathcal{C})$, yet the net $(\mathcal{D}, \tau)$ is not coherently (consistently) orientable, where $\mathcal{D} = VP \cup \{C_{k+1}\}$ and $\tau \subseteq \rho$. This conclusion is a contradiction to one of the hypotheses of the theorem.

We conclude that $(\{C_1, \ldots, C_{k+1}\}, \{\rho_1, \ldots, \rho_{k+1}\})$ is coherently (consistently) orientable. Thus, the proof is complete by induction. ■

Remark: For the search for conditions for the coherent or consistent orientability of $(\mathcal{D}, \tau)$, we may assume that no edge of G belongs to more than two elements of $\mathcal{D}$. This observation is immediate if $|\mathcal{D}| \geq 4$ (since $\mathcal{D}$ is the vertex set of an induced circuit). Suppose therefore that $\mathcal{D} = \{D_1, D_2, D_3\}$, where $D_1 \cap D_2 \cap D_3 \neq \emptyset$. Then $(\mathcal{D}, \tau)$ is not coherently orientable. Let $\tau = \{\tau_1, \tau_2, \tau_3\}$, where τ_i is an orientation of D_i for each $i \in J_3$. Choose $x \in D_1 \cap D_2 \cap D_3$ and $y \in D_1 \cap D_2$. We may assume without loss of generality that $\tau_1(x) = \tau_2(x) = \tau_3(x)$. Since $(\mathcal{D}, \tau)$ is pairwise coherently orientable, it follows that $\tau_1(y) = \tau_2(y)$. A similar argument obtains if $y \in (D_1 \cap D_3) \cup (D_2 \cap D_3)$. Thus $(\mathcal{D}, \tau)$ is consistently orientable.

The next theorem relates the coherent and consistent orientability of $(\mathcal{D}, \tau)$.

Theorem 4.2 *Let $(\mathcal{D}, \tau)$ be a pairwise coherently orientable net in G, where the graph $H(\mathcal{D})$ is connected and its edge set is a circuit. Suppose also that no edge of G belongs to more than two elements of $\mathcal{D}$.*

(1) *If $|\mathcal{D}|$ is even, then $(\mathcal{D}, \tau)$ is coherently orientable if and only if it is consistently orientable.*

(2) *Suppose that every set in $\mathcal{D}$ is a circuit, and that each orientation in τ is bijective. If $|\mathcal{D}|$ is odd, then $(\mathcal{D}, \tau)$ is coherently orientable if and only if it is not consistently orientable.*

Proof. Let $\mathcal{D} = \{D_1, \ldots, D_k\}$, where $D_i \cap D_{i+1} \neq \emptyset$ for each $i < k$. (Thus $D_1 \cap D_k \neq \emptyset$ also.) Let $\tau = \{\tau_1, \ldots, \tau_k\}$, where τ_i is an orientation of D_i for each i. We may assume that τ is a coherent (consistent) orientation of $\mathcal{D}$.

(1) Clearly, $\{\tau_1, \tau_2^*, \tau_3, \tau_4^*, \ldots, \tau_{k-1}, \tau_k^*\}$ is a consistent (coherent) orientation of $\mathcal{D}$.

(2) Let $\{\rho_1, \ldots, \rho_k\}$ be a consistent (coherent) orientation of $\mathcal{D}$, where ρ_i is an orientation of D_i for each i. Without loss of generality, we may take $\rho_1 = \tau_1$. Then $\rho_2 = \tau_2^*$, $\rho_3 = \tau_3$ and, since k is odd, we inductively obtain the contradiction that $\rho_1 = \tau_1^*$. Thus, if $(\mathcal{D}, \tau)$ is coherently orientable, then it cannot be consistently orientable. The converse follows because of pairwise coherent orientability. ■

4.3 Families of circuits

In this section, the results of the last are applied to the case where $\mathcal{C}$ is a set of circuits. Thus, we let $(\mathcal{C}, \rho)$ be a net $(\{C_1, \ldots, C_k\}, \{\rho_1, \ldots, \rho_k\})$, where ρ_i is an orientation of circuit C_i for each i.

If C is a circuit in a graph G, and σ is a bijective orientation of C, then we say that C is a *directed* circuit under σ. We assume throughout this section that C_i is a directed circuit under ρ_i for each i. In the light of Theorem 4.1, and using its terminology, we study the property of pairwise coherent orientability and the coherent or consistent orientability of $(\mathcal{D}, \tau)$.

If $\mathcal{C}$ is a set of circuits and ρ is an orientation of $\mathcal{C}$ whose elements are bijections, then the only other orientations of $\mathcal{C}$ with this property are those compatible with ρ. Moreover, if σ is such an orientation of $\mathcal{C}$, then $(\mathcal{C}, \sigma)$ is coherently (consistently) orientable if and only if $(\mathcal{C}, \rho)$ is coherently (consistently) orientable. We may therefore define $\mathcal{C}$ to be *coherently (consistently) orientable* if $(\mathcal{C}, \rho)$ is coherently (consistently) orientable.

Let C_1, C_2 be distinct circuits in G that meet. Let $S(C_1, C_2)$ be a graph whose vertices are those of degree 3 in $G[C_1 \cup C_2]$. Let the edges of $S(C_1, C_2)$ be the paths in G that join vertices of $S(C_1, C_2)$ and are either included in $C_1 \cap C_2$ or disjoint from $C_1 \cap C_2$ but included in C_1 or in C_2. Let each edge of $S(C_1, C_2)$ join the vertices of $S(C_1, C_2)$ that it joins in G. Thus, $S(C_1, C_2) = S(C_2, C_1)$. It is a cubic graph, called the *essence* of $\{C_1, C_2\}$.

Next, for each $i \in J_2$ let C_i' be the set of edges in $S(C_1, C_2)$ that are paths included in C_i in G. Then C_1' and C_2' are circuits in $S(C_1, C_2)$. Moreover, $\{C_1', C_2'\}$ is coherently orientable if and only if $\{C_1, C_2\}$ is coherently orientable.

We may therefore assume that $G[C_1 \cup C_2]$ is cubic, in which case we obtain the following theorem.

Theorem 4.3 *Let G be a cubic graph containing circuits C_1 and C_2 such that $EG = C_1 \cup C_2$. Then $\{C_1, C_2\}$ is coherently orientable if and only if G is bipartite.*

Proof. Suppose first that $\{C_1, C_2\}$ is coherently orientable. Let G be oriented so that C_1 and C_2 are directed circuits. Then every edge of $C_1 \cap C_2$ is directed from a vertex of indegree 2 to a vertex of indegree 1, and the remaining edges from a vertex of indegree 1 to one of indegree 2. Therefore, G is bipartite.

Conversely, suppose that G is bipartite, with bipartition $\{M, N\}$. Direct the edges of $C_1 \cap C_2$ from M to N, and the remaining edges from N to M. Then C_1 and C_2 are directed circuits. ■

We turn now to the set $\mathcal{D}$ of Theorem 4.1. Theorem 4.2 shows that it suffices to study the consistent orientability of $\mathcal{D}$. As in our study of pairwise coherent orientability, we show first that it suffices to investigate cubic graphs.

Let $\mathcal{C}$ be a set of circuits in a graph G. Let $a, b_1, b_2, \ldots, b_k$ be distinct edges of G incident on a common vertex v, where $k \geq 2$. We describe $(a, \{b_1, \ldots, b_k\})$ as a *fork* of $\mathcal{C}$ with *cusp* v if there are circuits $B_1, \ldots, B_k$ in $\mathcal{C}$ such that $\{a, b_i\} \subseteq B_i$ for each i, and k is maximal with respect to this property. We refer to a as the *handle* of the fork, and to $b_1, \ldots, b_k$ as the *tines*. A fork is said to *contain* its handle and its tines.

Example 4.3 *In Figure 4.3a, $\{C_1, C_2, C_3\}$ is a set $\mathcal{C}$ of circuits, where $C_1 = \{a, c, g\}$, $C_2 = \{a, b, d\}$, and $C_3 = \{e, f, g\}$. Clearly, $(a, \{d, g\})$ is a fork Λ_1 of $\mathcal{C}$ with handle a, cusp v, and tines d and g. Similarly, $(g, \{a, e\})$ is a fork Λ_2 with handle g, cusp v, and tines a and e.*

A path P in G is called *elementary* (with respect to $\mathcal{C}$) if:

(1) P is included in a member of $\mathcal{C}$,

(2) P joins cusps of forks, and

(3) P is minimal with respect to (2).

Let $A \in \mathcal{C}$ and $B \in \mathcal{C}$. Condition (3) implies that if $P \cap A \cap B \neq \emptyset$, then $P \subseteq A \cap B$, and that if $P \cap (A - B) \neq \emptyset$, then $P \subseteq A - B$.

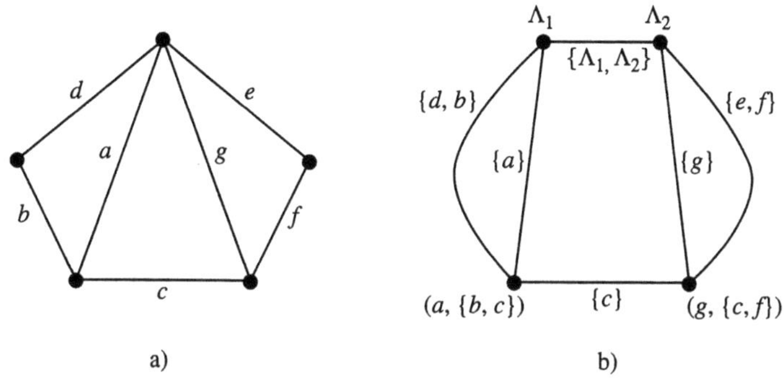

FIGURE 4.3.

Two forks of $\mathcal{C}$ with a common cusp v are said to be *locked* on A at v if A is a circuit of $\mathcal{C}$ containing both of their handles. Note that only two forks of $\mathcal{C}$ can be locked on A at v.

Example 4.4 *In Example 4.3, the forks Λ_1 and Λ_2 are locked on C_1 at v.*

Let $\mathcal{D}$ be the vertex set of an induced circuit C in $H(\mathcal{C})$, where no edge of G belongs to more than two circuits in $\mathcal{D}$. An edge of G is said to be *double* with respect to $\mathcal{D}$ if it carries two circuits of $\mathcal{D}$. Let $\mathcal{D} = \{C_0, \ldots, C_{l-1}\}$, where $C_i \cap C_{i+1} \neq \emptyset$ for each $i \in J^0_{l-2}$. Thus $C_0 \cap C_{l-1} \neq \emptyset$ since C is a circuit; but because C is induced, we have $C_i \cap C_j = \emptyset$ whenever $i \neq j$ and $i \not\equiv j \pm 1$. (All congruences and subscripts are to be read modulo l.) Since no edge of G carries more than two circuits of $\mathcal{D}$, any fork of $\mathcal{D}$ has only two tines. Moreover, two forks of $\mathcal{D}$ can be locked on only one circuit.

Let $S(\mathcal{D})$ be a graph whose vertices are the forks of $\mathcal{D}$ and whose edges are the elementary paths of G with respect to $\mathcal{D}$ and the pairs of forks of $\mathcal{D}$ locked on a common circuit of $\mathcal{D}$ at a common cusp. Each edge of $S(\mathcal{D})$ that is an elementary path P of G included in only one circuit in $\mathcal{D}$ is to join the vertices of $S(\mathcal{D})$ that contain a tine in P. Each edge of $S(\mathcal{D})$ that is an elementary path of G included in two circuits in $\mathcal{D}$ is to join the vertices of $S(\mathcal{D})$ whose handles it contains in G. Any other edge of $S(\mathcal{D})$ is to join its two members. We call $S(\mathcal{D})$ the *essence* of $\mathcal{D}$.

Example 4.5 *Figure 4.3b illustrates $S(\mathcal{C})$, where $\mathcal{C}$ is the set of circuits in Example 4.3.*

Let A be a circuit in $\mathcal{D}$. The forks of $\mathcal{D}$ with handles in A form the vertices of a circuit A' in $S(\mathcal{D})$ in which each edge is a path included in A or joins two forks locked on A. We describe A' as the *condensate* of A. An edge of A' joining two forks, F_1 and F_2, locked on A cannot belong to the condensate of any other circuit in $\mathcal{D}$, since A is the only circuit of $\mathcal{D}$ on which F_1 and F_2 are locked. It follows that $\mathcal{D}$ is a consistently orientable set of circuits in $\mathcal{C}$ if and only if the set of condensates of circuits in $\mathcal{D}$ is consistently orientable in $S(\mathcal{D})$. Moreover, every vertex of $S(\mathcal{D})$ is incident on an edge that carries only one condensate of a circuit in $\mathcal{D}$.

Lemma 4.4 *$S(\mathcal{D})$ is cubic.*

Proof. Any vertex F of $S(\mathcal{D})$ is a fork of $\mathcal{D}$. Let its cusp be v, its handle a, and its tines b_1 and b_2. Then $\mathcal{D}$ contains circuits C_i and C_{i+1} such that $\{a, b_1\} \subseteq C_i$ and $\{a, b_2\} \subseteq C_{i+1}$. Thus a is a double edge of G. We distinguish three cases, according to the number of tines of F that are double edges of G.

Case 1. Neither b_1 nor b_2 is double. Then each of a, b_1, b_2 is a terminal edge of an elementary path in G. These three elementary paths are the only edges of $S(\mathcal{D})$ incident on F.

Case 2. If only one of b_1, b_2 is double, we may suppose that b_1 is double and b_2 is not. C_{i-1} then contains b_1 and an edge b_0, which is distinct from a, b_1, b_2, and incident on v. Thus F and $(b_1, \{a, b_0\})$ are a pair of forks locked on C_i at v. An edge of $S(\mathcal{D})$ joins them. On the other hand, a and b_2 are terminal edges of two elementary paths in G that are edges of $S(\mathcal{D})$ incident on F. Hence F again has degree 3 in $S(\mathcal{D})$.

Case 3. Suppose that both b_1 and b_2 are double. Again C_{i-1} contains b_1 and another edge b_0 incident on v, and F and $(b_1, \{a, b_0\})$ are forks adjacent in $S(\mathcal{D})$. Moreover, C_{i+2} contains b_2 and another edge b_3 incident on v. The forks F and $(b_2, \{a, b_3\})$ are locked on C_{i+1} at v, and are hence adjacent in $S(\mathcal{D})$. Finally, a is a terminal edge of an elementary path in G which is an edge of $S(\mathcal{D})$ incident on F. Thus F has degree 3 in $S(\mathcal{D})$. ■

In view of the above considerations, we may now confine our attention to cubic graphs in which the set of double edges with respect to $\mathcal{D}$ is a 1-factor. We deduce the following theorem.

Theorem 4.5 *Let G be a cubic graph, and let $\mathcal{D}$ be a set of circuits in G such that $EG = \bigcup \mathcal{D}$, $\mathcal{D}$ is the vertex set of a circuit in $H(\mathcal{D})$, no edge of G belongs to more than two circuits of $\mathcal{D}$, and the set of double edges with respect to $\mathcal{D}$ is a 1-factor. Then $\mathcal{D}$ is consistently orientable if and only if G is bipartite.*

Proof. If $\mathcal{D}$ is consistently orientable, then G is bipartite by the argument in the proof of Theorem 4.3. Suppose therefore that G is bipartite, with bipartition $\{M, N\}$. Then the circuits of $\mathcal{D}$ are all directed if double edges are directed from M to N and the remaining edges from N to M. ■

As an application, we obtain the following theorem. When specialised to the case of a gem, it effects the reconciliation of our notion of an orientable imbedding with the traditional one.

Theorem 4.6 *Let $\mathcal{C}$ be the set of red-blue and blue-yellow bigons in a 3-graph K. Then $\mathcal{C}$ is consistently orientable if and only if K is bipartite.*

Proof. Suppose that K is bipartite. Let C, D be distinct bigons in $\mathcal{C}$ that meet. Then each blue edge of $K[C \cup D]$ joins two vertices of degree 2 in $K[C \cup D]$ or two vertices of degree 3 in $K[C \cup D]$. Since K is bipartite, it follows that the essence of $\{C, D\}$ is bipartite. We conclude from Theorem 4.3 that $\{C, D\}$ is coherently orientable. Thus $\mathcal{C}$ is pairwise coherently orientable.

Next, let $\mathcal{A}$ be the vertex set of an induced circuit in $H(\mathcal{C})$ and let $\mathcal{B}$ be the set of condensates of circuits in $\mathcal{A}$. Thus $H(\mathcal{B})$ has a single circuit, and the vertex set of this circuit is $VH(\mathcal{B}) = \mathcal{B}$. Moreover, no edge of the cubic graph $S(\mathcal{A})$ belongs to more than two circuits of $\mathcal{B}$, and $ES(\mathcal{A}) = \bigcup \mathcal{B}$. Since $S(\mathcal{A})$ is cubic and every vertex of $S(\mathcal{A})$ is incident on an edge that carries only one member of $\mathcal{B}$, it follows that the set of double edges of $S(\mathcal{A})$ with respect to $\mathcal{B}$ is a 1-factor of $S(\mathcal{A})$. In addition, $S(\mathcal{A})$ is bipartite by the argument used in the previous paragraph to show that the essence of $\{C, D\}$ is bipartite. Thus $\mathcal{B}$ is consistently orientable by Theorem 4.5. Hence $\mathcal{A}$ is consistently orientable. It follows by Theorem 4.1 that $\mathcal{C}$ is consistently orientable.

Conversely, let $\mathcal{C}$ be consistently orientable. Then K has an orientation under which every circuit of $\mathcal{C}$ is directed. Under this orientation, each edge of K joins a vertex of indegree 1 to a vertex of indegree 2. Hence K is bipartite. ■

4.4 Rings

The results of the previous sections are applicable to a characterisation of planar graphs proved in Chapter 9. A *ring* of circuits in a graph G is defined as a set $\mathcal{C}$ of consistently orientable circuits such that

(1) $|\mathcal{C}| \geq 3$;

(2) there is a cyclic permutation $(C_0\, C_1 \cdots C_{n-1})$ of the n circuits in $\mathcal{C}$ such that, for all i, C_i meets C_{i-1} and C_{i+1} but no other circuit of $\mathcal{C}$; and

(3) $C_{i-1} \cap C_i \cap C_{i+1} = \emptyset$ for all i.

[Here all subscripts are to be read modulo n. Note also that (2) implies (3) except when $n = 3$.] A ring of bonds is defined analogously. In both cases, an *n-ring* is a ring of cardinality n.

The ring $\mathcal{C}$ above is *odd* or *even* according to whether $|\mathcal{C}|$ is odd or even. We call $|\mathcal{C}|$ the *length* of the ring. Moreover, a ring $\mathcal{C}$ of circuits is *strict* if $|VC_i \cap VC_j| \leq 1$ whenever $C_i \cap C_j = \emptyset$. A ring $\mathcal{C}$ of bonds is *strict* if there do not exist distinct bonds A, B, C satisfying the conditions $B \in \mathcal{C}$, $C \in \mathcal{C}$, $B \cap C = \emptyset$, $A \subseteq B \cup C$. A ring $\{C_0, C_1, \ldots, C_{n-1}\}$ of bonds or circuits is *maximal* if there is no ring $\{C'_0, C'_1, \ldots, C'_{m-1}\}$ such that $\bigcup_{j=0}^{m-1} C'_j \subseteq \bigcup_{j=0}^{n-1} C_j$ and $m > n$.

Example 4.6 *In the graph shown in Figure 8.7b (p. 129), the set*

$$\{\{e_0, e_1, e_5\}, \{e_1, e_2, e_6\}, \{e_2, e_3.e_7\}, \{e_3, e_4, e_8\}, \{e_4, e_0, e_9\}\}$$

is a strict odd ring of circuits. It will be shown in Chapter 8 to be maximal.

The characterisation of planar graphs given in Chapter 9 asserts that a graph is planar if and only if it has no strict maximal odd ring of circuits. We obtain this result by dualising a theorem asserting that in any graph every strict maximal ring of bonds is even.

Let us define a set $\mathcal{C}$ of circuits in a graph G to be a *closed chain* if the requirement of consistent orientability in the definition of a ring is weakened to pairwise coherent orientability. Thus, every ring of circuits is a closed chain. Moreover, Theorem 4.2 implies that G is planar if and only if every consistently orientable, strict, maximal closed chain is also coherently orientable. Furthermore, suppose that G is cubic. Let $T(\mathcal{C})$ be a graph whose vertices are those of degree 3 in $G[\bigcup \mathcal{C}]$. Let the edges of $T(\mathcal{C})$ be the paths in $G[\bigcup \mathcal{C}]$, of minimal length, which join vertices of $T(\mathcal{C})$. Let each edge of $T(\mathcal{C})$ join the vertices of $T(\mathcal{C})$ that it joins in G. Thus $T(\mathcal{C})$ is a cubic graph. Theorem 4.5 then shows that G is planar if and only if there is no maximal odd closed chain $\mathcal{C}$ for which $T(\mathcal{C})$ is bipartite.

5
Non-separating Curves in Surfaces

An important theorem of topology asserts that the first Betti number of a surface is the largest number of closed curves that can be drawn in the surface without dividing it into two or more regions. We now generalise this theorem (which is itself a generalisation of the Jordan curve theorem) to 3-graphs. The topological implications of our results are discovered by specialising the main theorems to the case of gems.

5.1 The main results and their topological implications

Let K be a 3-graph. A non-empty set C of edges of K is called a *b-cycle* if C is the union of disjoint circuits with at least one blue edge in each. A set $\mathcal{C}$ of b-cycles *induces* a b-cycle C if each blue edge of C is an element of $\bigcup \mathcal{C}$. In the special case where $\mathcal{C} = \{D\}$ for a b-cycle D, we also say that D *induces* C. For example, any b-cycle induces itself. The *boundary space* $\mathcal{F}(K)$ of K is the subspace of $\mathcal{C}(K)$ spanned by the set of bigons of K. A b-cycle is said to *separate* if it induces a b-cycle which is a member of the boundary space. A set $\mathcal{C}$ of b-cycles is said to *separate* if it induces a b-cycle that separates. A b-cycle is *connected* if it is a circuit.

Example 5.1 *Consider the 3-graph of Figure 5.1 and let* $C = \{b_1, c_3, b_3, c_2, a_1\}$. *The b-cycle C separates since it induces the b-cycle* $\{b_3, a_3\}$. *However, C is not a member of the boundary space since it is not a sum of bigons.*

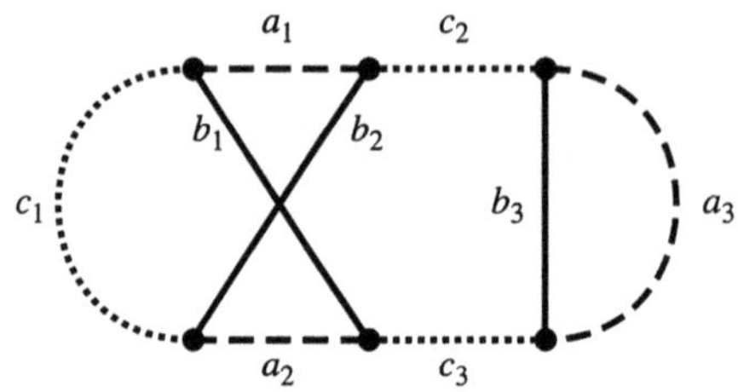

FIGURE 5.1.

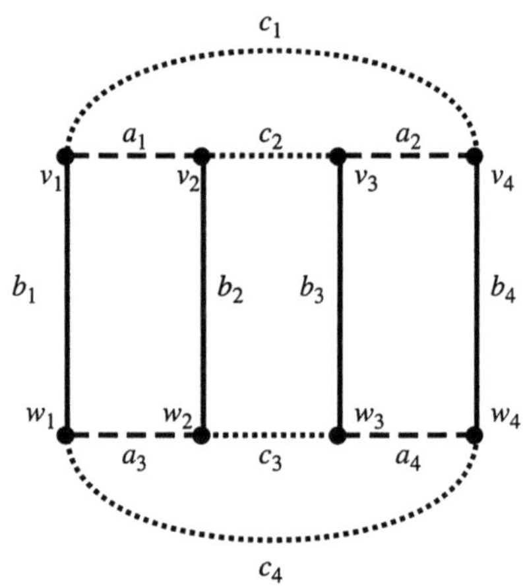

FIGURE 5.2.

If C is a b-cycle, its *necklace* $N(C)$ is the set of red-yellow bigons that it meets. The elements of its necklace are the *beads* of C. The *poles* of a bead B (with respect to C) are the vertices of B incident on a blue edge of C. If C is connected and each bead has just two poles, then C is a *semicycle*.

Example 5.2 *Consider the 3-graph of Figure 5.2 and let C be the connected b-cycle $\{b_2, c_2, a_2, b_4, a_4, c_3\}$. Let $B_1 = \{a_1, c_2, a_2, c_1\}$ and $B_2 = \{a_3, c_3, a_4, c_4\}$. Then $N(C) = \{B_1, B_2\}$. The poles of B_1 with respect to C are v_2 and v_4. Likewise, the poles of B_2 with respect to C are w_2 and w_4. Hence C is a semicycle. The connected b-cycle $\{a_1, b_2, c_3, b_3, a_2, b_4, c_4, b_1\}$ is not a semicycle since the poles of B_1 are v_1, v_2, v_3, and v_4.*

The concept of a semicycle was introduced by Stahl (1983) in the setting of permutation pairs, but the motivation for it is best explained by considering the case where K is a gem. As we indicated in Section 2.4, K then corresponds to an imbedding, in a closed surface $\mathcal{S}$, of a graph G. Under this interpretation, the beads of a semicycle C of K correspond to vertices of G. The requirements that C should be connected and have a blue edge, and that each bead should have just two poles, reveal that C corresponds to a circuit D of G or a path of length 1. In the former case, observe that if D divides $\mathcal{S}$ into two regions, then D is the sum of the boundaries of the faces inside one of those regions. If R is either of those regions, then C is the sum of the blue-yellow bigons of K corresponding to the faces inside R, the red-yellow bigons corresponding to vertices in the interior of R, the red-blue bigons corresponding to edges in the interior of R, and possibly some of the red-yellow and red-blue bigons corresponding to the vertices and edges, respectively, of D. We infer that C is a member of the boundary space, and hence separates. Conversely, if the semicycle C separates, then C is a sum of bigons, and it follows that D divides $\mathcal{S}$ into two regions. The vertices, edges, and faces in the interior of one of these regions correspond to those bigons in the sum that are not beads of C or red-blue bigons that meet C.

On the other hand, suppose that C corresponds to a path Q in G of length 1. Then C has just two beads. Either C is the red-blue bigon corresponding to the unique edge of Q, or C is the sum of this bigon and one or both of the beads. In any case, C separates.

Note also that each circuit of G has a non-empty set of semicycles of K that correspond to it.

The members of a set $\mathcal{C}$ of b-cycles are *b-independent* if each b-cycle in $\mathcal{C}$ contains a blue edge that is not in any of the others. The b-cycles in $\mathcal{C}$ are *b-dependent* if they are not b-independent. Hence, the members of $\mathcal{C}$ are b-dependent if there exists $C \in \mathcal{C}$ induced by $\mathcal{C} - \{C\}$.

Example 5.3 *Consider the gem of Figure 5.3. Let* $C_1 = \{b_5, c_2, a_4, c_3, b_2, c_4\}$ *and* $C_2 = \{b_4, c_6, b_1, c_1, a_3, c_2\}$. *Hence, if* $\mathcal{C} = \{C_1, C_2\}$, *then* $\mathcal{C}$ *is a set of two b-independent semicycles, since the blue edge* $b_4 \in C_2$ *is not in* C_1 *and, similarly, the blue edge* $b_5 \in C_1$ *is not in* C_2. *Now let* $C_3 = \{a_1, b_2, a_2, b_1\}$. *Then* $\{C_1, C_2, C_3\}$ *is a set of b-dependent semicycles since* C_3 *is induced by* $\{C_1, C_2\}$.

The following theorem will be proved in Section 5.4.

Theorem 5.1 *If* K *is a* 3*-graph, then the cardinality of a maximum set of b-independent semicycles that does not separate is* $2c(K) - \chi(K)$.

In the remainder of this section, we explain the topological significance of this result and show how it can be used to deduce the version of the Jordan curve theorem that appeared in Stahl (1983). For convenience, we call $2c(K) - \chi(K)$ the *first Betti number* of K.

The *conjugate* C^* of a b-cycle C is defined as $C + (\bigcup N(C))$ and is also a b-cycle. We define $I(C) = C \cap (\bigcup N(C))$ and $O(C) = C^* \cap (\bigcup N(C))$. We note that $\{I(C), O(C)\}$ is a partition of $\bigcup N(C)$. We call $I(C)$ and $O(C)$ the *sides* of C. This terminology is suggested by the interpretation of a gem as a model for an imbedding of a graph. Note that C^* separates if and only if C does.

Example 5.4 *Consider* C_1 *and* C_2 *of Example 5.3. Then* $I(C_1) = \{c_2, a_4, c_3, c_4\}$ *and* $O(C_1) = \{a_3, c_1, a_1, a_2, c_6, a_6, c_5, a_5\}$. *Similarly,* $I(C_2) = \{c_6, c_1, a_3, c_2\}$ *and* $O(C_2) = \{a_6, c_5, a_5, c_4, a_2, a_1, c_3, a_4\}$.

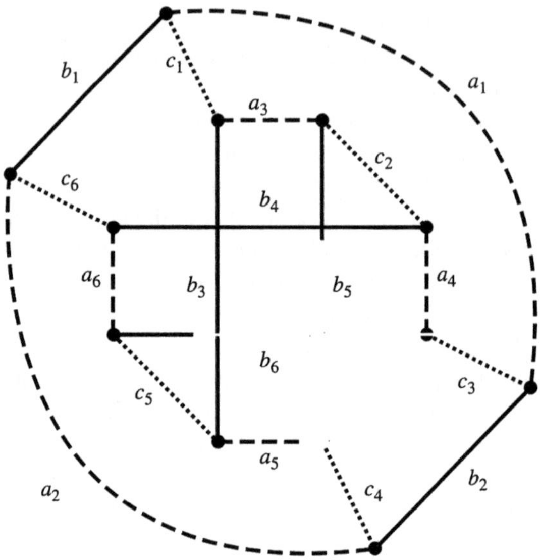

FIGURE 5.3.

Now let P be a non-empty path in K whose terminal edges are blue. For such paths, we define *necklaces*, *beads*, *poles*, and *conjugates* as for b-cycles, and we use analogous notation for these concepts. If each bead of the necklace $N(P)$ has just two poles, we call P a *semipath* of K. In addition, a red-yellow bigon is called a *terminal bead* of P if it contains a terminal vertex of P. P therefore has just one or two terminal beads. The elements of $N(P)$ are sometimes called *internal beads* of P. Each internal bead of a semipath has just two poles, and so none of them is terminal.

Example 5.5 *Consider the gem of Figure 5.3 and let* $P = \{b_1, c_6, b_4\}$. *Then* P *is a semipath with one terminal bead* $\{c_1, a_3, c_2, a_4, c_3, a_1\}$ *and one internal bead* $\{c_6, a_6, c_5, a_5, c_4, a_2\}$.

If K is a gem representing an imbedding of a graph G, then a semipath P represents a path or circuit in G according to whether P has two terminal beads or just one. If P represents a path, then the internal and terminal beads of P correspond to the internal and terminal vertices, respectively, of the path. If P represents a circuit, then the internal beads and the unique terminal bead correspond to the vertices of the circuit.

A b-cycle C and a semipath P are said to *miss* if they are disjoint and no internal bead of P is a bead of C. [If K is a gem, C a semicycle in K corresponding to a circuit D in $G(K)$, and P a semipath in K corresponding to a path Q in $G(K)$, then C and P miss if and only if D and Q are disjoint and no internal vertex of Q is a vertex of D. If Q is a circuit rather than a path, then C and P miss if and only if D and Q are disjoint and no vertex of Q corresponding to some internal bead of P is a vertex of D.] If C and P miss, then P is said to *link the sides* of C if one terminal vertex is in $VI(C)$ and the other is in $VO(C)$. (We note that the terminal vertices of P cannot be poles of a bead of C since $P \cap C = \emptyset$.)

Example 5.6 *The semipath $\{b_6\}$ links the sides of the semicycle C_1 of Example 5.3.*

A b-cycle C is *normal* if no blue edge of C is adjacent to edges of C with distinct colours. Clearly, the conjugate of a normal b-cycle is also normal. Furthermore, no b-cycle of odd length can be normal.

The following theorem is proved in Section 5.3.

Theorem 5.2 *A necessary and sufficient condition for a b-cycle C to separate a 3-graph K is for C to induce a normal b-cycle C' such that no semipath links the sides of C'.*

Theorem 5.2 asserts that a b-cycle separates if and only if it induces a normal b-cycle whose sides are not linked by a semipath. This theorem also has a topological interpretation, which may be discerned from the following considerations. Let K be a gem corresponding to an imbedding of a graph G in a closed surface $\mathcal{S}$, and let D be a circuit in G corresponding to a normal semicycle C in K. Then the two sides of C determine a pair of complementary subsets E_1 and E_2 of the set $\partial_G VD$. Each of these subsets consists of all of the edges in $\partial_G VD$ whose corresponding red-blue bigons in K meet a given side of C. Intuitively, E_1 and E_2 may be thought of as representing the sides of D in the imbedding of G, since C is normal. Now let P be a semipath in K of length greater than 1. If P corresponds to a path Q in G, then P links the sides of C in K if and only if Q joins two vertices of D, meets both E_1 and E_2, and none of its internal vertices is a vertex of D. On the other hand, if Q is a circuit rather than a path, then P links the sides of C if and only if Q meets both E_1 and E_2 and has just one vertex in common with D.

Intuitively, Theorem 5.2 therefore asserts that C separates if and only if D divides $\mathcal{S}$ into two regions in such a way that if Q is a path in G with no internal vertex in VD, or a circuit in G having, at most, one vertex in VD, then the vertices and edges of Q that are not in D are collectively confined to one region.

Example 5.7 *Consider the b-cycle C of Example 5.1. C induces the normal b-cycle $C' = \{b_3, a_3\}$. Moreover, $I(C') = \{a_3\}$ and $O(C') = \{c_3, a_2, c_1, a_1, c_2\}$, and therefore it is evident that no semipath links the sides of C'. This result, of course, agrees with Theorem 5.2.*

If T is either a set of edges of a 3-graph K or a subgraph of K, we denote by $\beta(T)$ the set of all blue edges of T. If C is a semicycle of K that induces another b-cycle C', then C' is a semicycle such that $\beta(C') = \beta(C)$. Now suppose that K is a gem corresponding to an imbedding of a connected graph G in a closed surface $\mathcal{S}$. It follows from the above observation that C and C' correspond to the same path or circuit of G. Note also that a set of semicycles in K is b-independent if and only if the corresponding paths and circuits in G have the property that each contains an edge not in the union of the others. Thus, a necessary and sufficient condition for a set of semicycles to be a set $\mathcal{C}$ of b-independent semicycles that does not separate

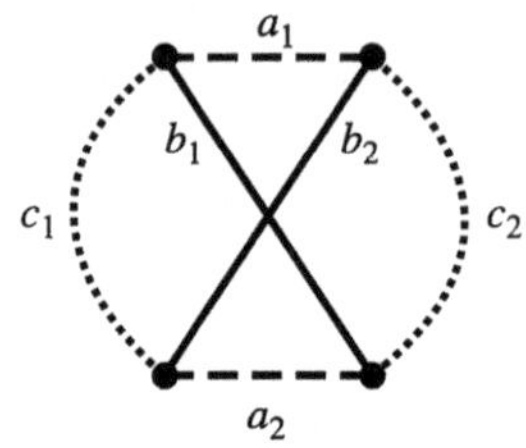

FIGURE 5.4.

is for C to correspond to a set of circuits in G that are collectively drawn so as not to divide $\mathcal{S}$ into two or more regions and have the property that each contains an edge not in any of the others. According to Theorem 5.1, the cardinality of a maximum set of such semicycles is $2 - \chi(\mathcal{S})$, since $c(K) = c(G) = 1$. This number is the first Betti number of $\mathcal{S}$. These observations show that Theorem 5.1 is a graph theoretic version of the theorem that the first Betti number of a surface is the largest number of closed curves that can be drawn in the surface without dividing it into two or more regions.

In order to obtain a graph theoretic version of the Jordan curve theorem, let us suppose first that $\mathcal{S}$ is the sphere. Since the sphere is orientable, K is bipartite. Let C be a normal semicycle in K. By Theorem 5.1, $\{C\}$ separates and therefore induces a b-cycle that separates. By Theorem 5.2, this b-cycle induces a normal b-cycle D whose sides are not linked by a semipath. D therefore separates. But D, being normal and induced by $\{C\}$, must be C or C^*. In either case we conclude that C separates.

On the other hand, suppose that K is bipartite but that $\mathcal{S}$ is not the sphere. Then K has a semicycle C that does not separate. C therefore does not induce a b-cycle that is a sum of bigons. But it is easy to show that C induces a normal semicycle. We begin by observing that if b is a blue edge of C adjacent to edges of C with distinct colours, then by adding to C a bead containing one of these edges we obtain a semicycle induced by C in which the edges adjacent to b are of the same colour. By applying this construction to all but one of the blue edges of C, we can find a semicycle C' that is induced by C and has the property that, at most, one blue edge of C' is adjacent to edges of C' with distinct colours. The requirement that $|C'|$ be even, because K is bipartite, then forces C' to be normal. Since C' is induced by C, it cannot be a sum of bigons. Any b-cycle induced by C' is obtained from C' by the addition of beads of C', and therefore is not a sum of bigons either. We conclude that C' does not separate.

In summary, we have shown that if K is bipartite, then G is planar if and only if every normal semicycle of K separates. This is the graph theoretic version of the Jordan curve theorem that appeared in Little (1988). It implies another version, due to Stahl (1983), as we explain in the next section.

In the case where K is non-bipartite, it may not be possible to construct a normal semicycle, separating or non-separating. Consider the gem in Figure 5.4. The four semicycles in this gem are $\{b_1, c_1, a_2\}$ and $\{b_2, c_2, a_2\}$ and their conjugates. In all cases, the semicycle is not normal.

It follows from Lemma 1.35 (p. 20) that the dimension of the boundary space $\mathcal{F}(K)$ of a 3-graph K is $r(K) - c(K)$. Hence

$$\begin{aligned} 2c(K) - \chi(K) &= 2c(K) - r(K) + \frac{|VK|}{2} \\ &= |EK| - |VK| + c(K) - (r(K) - c(K)) \\ &= \dim \mathcal{C}(K) - \dim \mathcal{F}(K), \end{aligned}$$

since K is cubic. Thus, we obtain the following theorem.

Theorem 5.3 *A 3-graph K is spherical if and only if its bigons span $\mathcal{C}(K)$.*

We now deduce a consequence for imbeddings of graphs.

Theorem 5.4 *An imbedding I of a graph G is planar if and only if its face boundaries span $\mathcal{C}(G)$.*

Proof. Suppose that I is planar. Let K be the gem that encodes I. Any circuit C in G corresponds to a semicycle D of K. As K is spherical, we may write $D = \sum \mathcal{U}$ for some set $\mathcal{U}$ of bigons of K. Let $\mathcal{R}$ be the set of blue-yellow bigons in $\mathcal{U}$. Each member of $\mathcal{R}$ corresponds to a face of I, and the sum of the boundaries of these faces yields C.

Conversely, if I is not a planar imbedding of G, then, by Corollary 3.10 (p. 51), its face boundaries do not span $\mathcal{C}(G)$ as they are linearly dependent. ■

5.2 Permutation pairs

The concept of a *permutation pair*, an ordered pair of permutations of a set S, has been studied in several papers Stahl (1980, 1983, 1988). Each permutation pair (α, γ) corresponds to a bipartite 3-graph $K(\alpha, \gamma)$. The vertex set of this graph is the Cartesian product $S \times \{-1, 1\}$. For each $a \in S$, $K(\alpha, \gamma)$ contains a red edge joining $(a, 1)$ and $(a, -1)$, a blue edge joining $(a, 1)$ and $(\gamma a, -1)$, and a yellow edge joining $(a, -1)$ and $(\alpha a, 1)$. (See Figure 5.5.) Conversely, each bipartite 3-graph represents a permutation pair.

Example 5.8 *Let $S = J_9$. Then (α, γ) is a permutation pair, where $\alpha = (123)(456)(789)$ and $\gamma = (23)(567)(48)$. This is, in fact, Example 5.1 of Stahl (1983). Its corresponding 3-graph is given in Figure 5.6.*

Let (α, γ) be a permutation pair acting on a set S. The orbits of α are in 1:1 correspondence with the red-yellow bigons of $K(\alpha, \gamma)$ and the orbits of γ with the red-blue bigons of $K(\alpha, \gamma)$. Inspection of Figure 5.5 also reveals the existence of a similar 1:1 correspondence between the orbits of $\alpha\gamma$ and the blue-yellow bigons of $K(\alpha, \gamma)$. If γ is an involution, then the red-blue bigons are squares, so that

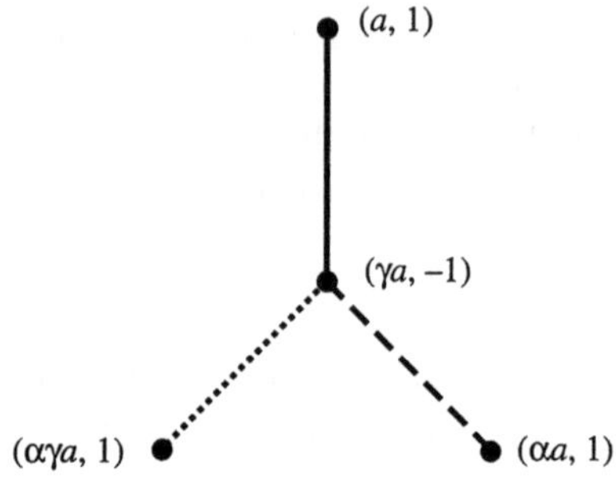

FIGURE 5.5.

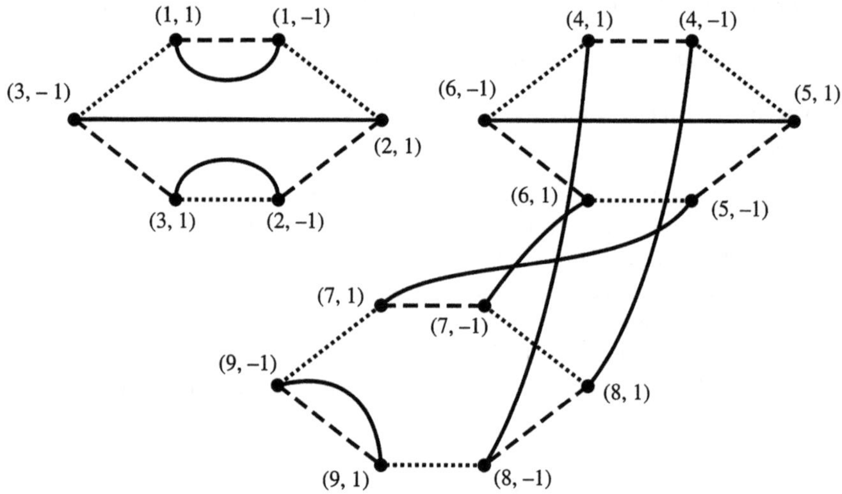

FIGURE 5.6.

$K(\alpha, \gamma)$ is a gem. Thus, the notion of a permutation pair generalises the concept of an imbedding of a graph in an orientable surface. Note also that each $a \in S$ corresponds to a red edge $a' \in EK(\alpha, \gamma)$.

Stahl (1983) defines an *xy-semipath* D in (α, γ) as an ordered set

$$(x, a_1, \alpha_1, b_1, a_2, \alpha_2, b_2, \ldots, a_t, \alpha_t, b_t, y)$$

for which

(1) $\alpha_1, \ldots, \alpha_t$ are distinct orbits of α,

(2) $a_i, b_i \in \alpha_i$ for each i, and

(3) there exist $\varepsilon_0, \varepsilon_1, \ldots, \varepsilon_t \in \{1, -1\}$ satisfying the condition that $\gamma^{\varepsilon_i} b_i = a_{i+1}$ for each $i \in J_t^0$, where $b_0 = x$ and $a_{t+1} = y$.

Example 5.9 *The ordered set*

$$(8, 4, (456), 6, 7, (789), 8, 4)$$

gives an 84*-semipath D in Example 5.8.*

Likewise, a *semicycle* C in (α, γ) is an ordered set

$$(a_1, \alpha_1, b_1, a_2, \alpha_2, b_2, \dots, a_t, \alpha_t, b_t),$$

where $t > 0$, for which (1) and (2) hold and there exist $\varepsilon_1, \varepsilon_2, \dots, \varepsilon_t \in \{1, -1\}$ such that $\gamma^{\varepsilon_i} b_i = a_{i+1}$ for each $i \in J_t$, where $a_{t+1} = a_1$.

Example 5.10 *The ordered set*

$$(4, (456), 6, 7, (789), 8)$$

gives a semicycle in Example 5.8.

The orbits $\alpha_1, \alpha_2, \dots, \alpha_t$ are the *vertices* of D and C in (α, γ). The *arcs* of D are the ordered pairs $(x, a_1), (b_1, a_2), (b_2, a_3), \dots, (b_{t-1}, a_t), (b_t, y)$, and the *arcs* of C are $(b_1, a_2), (b_2, a_3), \dots, (b_{t-1}, a_t), (b_t, a_1)$.

If $t > 1$ and neither x nor y appears in any of the orbits $\alpha_1, \dots, \alpha_t$, then the xy-semipath D in (α, γ) given above corresponds in the following way to a collection $\mathcal{D}$ of semipaths in $K(\alpha, \gamma)$ such that the internal beads of each member of $\mathcal{D}$ are the red-yellow bigons $B_1', B_2', \dots, B_t'$ of $K(\alpha, \gamma)$ corresponding to $\alpha_1, \alpha_2, \dots, \alpha_t$, respectively. First, condition (3) shows that for each arc there is a corresponding blue edge of $K(\alpha, \gamma)$ adjacent to the red edges corresponding to the components of the arc. (This blue edge is not necessarily unique.) For each member of $\mathcal{D}$, the set of blue edges is obtained by selecting one such blue edge for each arc. Let us refer to a vertex as a pole if it is incident on one of the selected blue edges. For each α_i there exist paths in B_i' joining poles. Condition (1) implies that for each i there are exactly two choices for such a path, the choices being complementary subsets of B_i'. The red and yellow edges of a member of $\mathcal{D}$ are supplied by choosing such a path for each i. It is immediate that each member of $\mathcal{D}$ is a semipath in $K(\alpha, \gamma)$.

In a similar way, we find that if $t > 2$, then each semicycle

$$(a_1, \alpha_1, b_1, \dots, a_t, \alpha_t, b_t)$$

in (α, γ) corresponds to a collection of semicycles in $K(\alpha, \gamma)$. The beads of each member of this collection are the red-yellow bigons $B_1', B_2', \dots, B_t'$ of $K(\alpha, \gamma)$ corresponding to $\alpha_1, \alpha_2, \dots, \alpha_t$, respectively. For each arc, every member of the collection contains a corresponding blue edge adjacent in $K(\alpha, \gamma)$ to the red edges corresponding to the components of the arc. Since $K(\alpha, \gamma)$ is bipartite, we find that for each semicycle in (α, γ) the corresponding collection of semicycles in $K(\alpha, \gamma)$ contains a normal semicycle.

Example 5.11 *In Example 5.8, Stahl defines the semicycle*

$$C = (4, (456), 6, 7, (789), 8)$$

in (α, γ)*. An example of a normal semicycle in* $K(\alpha, \gamma)$ *corresponding to* C *is the one with vertex set*

$$\{(4, -1), (5, 1), (5, -1), (6, 1), (7, -1), (8, 1)\}.$$

Note that the semipath (x, y) in (α, γ) corresponds in $K(\alpha, \gamma)$ to a semipath consisting of a single blue edge.

Next, suppose that D is the semipath $(x, a_1, \alpha_1, b_1, y)$ in (α, γ), where neither x nor y appears in the orbit α_1. If either $a_1 \neq b_1$ or $x \neq y$, then D corresponds to a collection of semipaths in $K(\alpha, \gamma)$ by the construction given above. If $a_1 = b_1$ and $x = y$, then D still corresponds to a collection of semipaths in $K(\alpha, \gamma)$ provided that (xa_1) is an orbit of γ, because in that case there are two blue edges in $K(\alpha, \gamma)$ adjacent to both x' and a_1'. However, if $a_1 = b_1$, $x = y$, and (xa_1) is not an orbit of γ, then $(x, a_1) = (y, b_1)$ but only one blue edge of $K(\alpha, \gamma)$ is adjacent to x' and a_1'. Only under these circumstances does D not correspond to a collection of semipaths in $K(\alpha, \gamma)$. In this case, we describe D as *singular*.

Any semicycle in (α, γ) of the form (a_1, α_1, b_1) corresponds to a collection of semicycles in $K(\alpha, \gamma)$ by the construction given above. Stahl defines a semicycle of the form (a_1, α_1, a_1) to be *degenerate*. The collection of corresponding semicycles in $K(\alpha, \gamma)$ consists of a red-blue digon and its conjugate.

Now consider a semicycle in (α, γ) of the form $(a_1, \alpha_1, b_1, a_2, \alpha_2, b_2)$. By the argument used above for the semipath $(x, a_1, \alpha_1, b_1, y)$, this semicycle corresponds to a collection of semicycles in $K(\alpha, \gamma)$ except in the case where $a_1 = b_1, a_2 = b_2$, and (a_1a_2) is not an orbit of γ. Stahl also defines a semicycle of the form $(a_1, \alpha_1, a_1, a_2, \alpha_2, a_2)$ to be *degenerate*, and *singular* if in addition (a_1a_2) is not an orbit of γ. Thus, a singular semicycle in (α, γ) has no corresponding collection of semicycles in $K(\alpha, \gamma)$.

Example 5.12 *Stahl illustrates these ideas in Example 5.8 with the following semicycles:*

$$\begin{aligned}
C_1 &= (1, (123), 1), \\
C_2 &= (2, (123), 3), \\
C_3 &= (5, (456), 6), \\
C_4 &= (4, (456), 4, 8, (789), 8), \\
C_5 &= (6, (456), 6, 7, (789), 7), \\
C_6 &= (4, (456), 6, 7, (789), 8).
\end{aligned}$$

Of these, only C_5 *is singular. In Figure 5.6 the semicycles with vertex sets*

$$\{(1, 1), (1, -1)\},$$

$$\{(3,1),(2,-1)\},$$
$$\{(5,1),(6,-1),(4,1),(4,-1)\},$$
$$\{(4,1),(8,-1),(8,1),(4,-1)\},$$
$$\{(4,-1),(4,1),(6,-1),(6,1),(7,-1),(8,1)\}$$

are members of the families corresponding to C_1, C_2, C_3, C_4, C_6, *respectively.*

Note also that every semipath in $K(\alpha,\gamma)$ corresponds to a semipath in (α,γ), and every semicycle in $K(\alpha,\gamma)$ corresponds to a semicycle in (α,γ). Moreover, the vertices of a semipath or semicycle in (α,γ) correspond to the (internal) beads of each member of the corresponding collection of semipaths or semicycles in $K(\alpha,\gamma)$.

A semipath and a semicycle in (α,γ) are said to *intersect* if they share a vertex. Corresponding semipaths and semicycles in $K(\alpha,\gamma)$ therefore do not miss.

An arc (b,a) of a semipath D or semicycle C in (α,γ) is said to be *coherent* if $a=\gamma b\neq\gamma^{-1}b$, *retrograde* if $a=\gamma^{-1}b\neq\gamma b$, and *neutral* if $a=\gamma b=\gamma^{-1}b$. If C or D corresponds to a collection of semicycles or semipaths in $K(\alpha,\gamma)$, then in any member of this collection the blue edges corresponding to neutral arcs are those that belong to red-blue squares of $K(\alpha,\gamma)$. Note also that an arc (b,a) in C or D corresponds to a blue edge joining vertices $(b,1)$ and $(a,-1)$ if the arc is coherent and to a blue edge joining vertices $(b,-1)$ and $(a,1)$ if the arc is retrograde.

We now come to the concept of the sides of a non-degenerate semicycle. Let C be the non-degenerate semicycle

$$(a_1,\alpha_1,b_1,a_2,\alpha_2,b_2,\ldots,a_t,\alpha_t,b_t).$$

Choose $i\in J_t$. If $a_i\neq b_i$, define $L(\alpha_i)=\{\alpha a_i,\alpha^2a_i,\ldots,\alpha^{k-1}a_i\}$, where k is the smallest positive integer for which $\alpha^k a_i=b_i$. Suppose now that $a_i=b_i$. Since C is non-degenerate, it follows that $t>1$ and that the arc (b_{i-1},a_i) cannot be neutral. (Here, and in the following, we write $b_0=b_t$ and $a_{t+1}=a_1$.) Suppose that this arc is coherent. Then $a_i=\gamma b_{i-1}$. If $a_{i+1}=\gamma^{-1}b_i$, then $a_{i+1}=\gamma^{-1}a_i=b_{i-1}$, in contradiction to the assumption that C is non-degenerate. We infer that if the arc (b_{i-1},a_i) is coherent, then so is (b_i,a_{i+1}). Similarly, if the former arc is retrograde, then so is the latter. We define $L(\alpha_i)=\alpha_i-\{a_i\}$ if (b_{i-1},a_i) is coherent, and $L(\alpha_i)=\emptyset$ if (b_{i-1},a_i) is retrograde. In both cases, there is a unique blue edge of $K(\alpha,\gamma)$ corresponding to the arc (b_{i-1},a_i). In the former case, this blue edge joins the vertices $(b_{i-1},1)$ and $(a_i,-1)$; in the latter case, it is the vertices $(b_{i-1},-1)$ and $(a_i,1)$ that are so joined.

In $K(\alpha,\gamma)$, let $\rho(\alpha_i)$ be the set of red edges corresponding to elements of $L(\alpha_i)$. Let B_i' be the red-yellow bigon in $K(\alpha,\gamma)$ corresponding to α_i. Let u_i and v_i be the vertices of B_i' incident on the blue edges corresponding to the arcs (b_{i-1},a_i) and (b_i,a_{i+1}), respectively. [If the arc (b_{i-1},a_i) is neutral, then the corresponding blue edge is not unique. In this case, we define $u_i=(a_i,-1)$. Similarly, if (b_i,a_{i+1}) is neutral, then we define $v_i=(b_i,1)$.] Let $I(B_i')$ be the path included in B_i' that joins

u_i and v_i and contains, as a terminal edge, the red edge incident on u_i if (b_{i-1}, a_i) is retrograde but not if that arc is coherent or neutral. Then $I(B_i')$ contains the red edge incident on v_i if and only if the arc (b_i, a_{i+1}) is retrograde. Moreover, the set of internal red edges of $I(B_i')$ is $\rho(\alpha_i)$.

Next, let $\beta(C)$ denote a set of blue edges of $K(\alpha, \gamma)$ corresponding to the arcs of C and chosen in such a way that $\bigcup_{i=1}^{t} I(B_i') \cup \beta(C)$ is a semicycle C' in $K(\alpha, \gamma)$ corresponding to C. Then, the beads of C' are $B_1', B_2', \ldots, B_t'$, and $I(C') = \bigcup_{i=1}^{t} I(B_i')$. Moreover, by the considerations of the previous paragraph, we see that C' is normal because a blue edge of C' corresponding to a retrograde arc is adjacent in C' only to red edges, and a blue edge corresponding to a coherent or neutral arc only to yellow edges.

Define $L(C)$ to be the union of $\bigcup_{i=1}^{t} L(\alpha_i)$ with all of the sets of the form $\{b_{i-1}, a_i\}$ where (b_{i-1}, a_i) is a retrograde arc. $L(C)$ therefore corresponds in $K(\alpha, \gamma)$ to the set of red edges of $I(C')$.

Similarly, for each i define $R(\alpha_i) = \alpha_i - (L(\alpha_i) \cup \{a_i, b_i\})$. Then $R(C)$ is defined as the union of $\bigcup_{i=1}^{t} R(\alpha_i)$ with all of the sets of the form $\{b_{i-1}, a_i\}$, where (b_{i-1}, a_i) is a coherent arc. The set of corresponding red edges in $K(\alpha, \gamma)$ is the set of red edges of $O(C')$ that do not belong to a red-blue square that contains a blue edge of C'. The *sides* of C are $L(C)$ and $R(C)$.

Example 5.13 *For the semicycle C of Example 5.11, we have $L(C) = \{5\}$ and $R(C) = \{6, 7, 9\}$.*

An xy-semipath D is said by Stahl to *link the sides* of a non-degenerate semicycle C if $x \in L(C)$, $y \in R(C)$, and C and D do not intersect. Thus C and D have no vertex in common. Therefore, neither x nor y appears in a vertex of D. Moreover, $x \neq y$. Hence D corresponds to a collection of semipaths in $K(\alpha, \gamma)$. Let D' be a member of this collection. Since C is non-degenerate, C corresponds to a collection of semicycles in $K(\alpha, \gamma)$. Let C' be a normal member of this collection. Without loss of generality, we may assume that $L(C)$ corresponds in $K(\alpha, \gamma)$ to a subset of $I(C')$, and $R(C)$ to a subset of $O(C')$. Moreover, the terminal edges of D' are adjacent in $K(\alpha, \gamma)$ to the red edge $x' \in I(C')$ and to the red edge $y' \in O(C')$, respectively. Note that if b is a blue edge of C', then b cannot be adjacent to a red edge of $I(C')$ and to a red edge of $O(C')$, since C' is normal. Hence $D' \neq \{b\}$, and since no internal bead of D' is a bead of C' it follows that C' and D' are disjoint and therefore miss. Hence D' links the sides of C' in $K(\alpha, \gamma)$. Conversely, if D' links the sides of C', then D links the sides of C.

Stahl defines a semicycle C of (α, γ) to *separate* if no semipath links its sides. Thus, every degenerate semicycle separates. A non-degenerate semicycle separates (α, γ) if and only if a corresponding normal semicycle separates $K(\alpha, \gamma)$.

The number of orbits of a permutation α is denoted by $\|\alpha\|$. The *genus* $g(\alpha, \gamma)$ of a permutation pair (α, γ) acting on a set S is defined as

$$c(\alpha, \gamma) - \frac{\|\alpha\| + \|\gamma\| + \|\alpha\gamma\| - |S|}{2},$$

where $c(\alpha, \gamma)$ is the number of orbits of the group generated by α and γ. In other words, $c(\alpha, \gamma)$ is the number of components of $K(\alpha, \gamma)$. Since $|VK(\alpha, \gamma)| = 2|S|$, we have

$$\begin{aligned} g(\alpha, \gamma) &= c(K(\alpha, \gamma)) - \frac{r(K(\alpha, \gamma))}{2} + \frac{|VK(\alpha, \gamma)|}{4} \\ &= c(K(\alpha, \gamma)) - \frac{\chi(K(\alpha, \gamma))}{2}, \end{aligned}$$

half the first Betti number of $K(\alpha, \gamma)$.

Our version of the Jordan curve theorem therefore asserts that every normal semicycle of $K(\alpha, \gamma)$ separates if and only if $g(\alpha, \gamma) = 0$. Stahl's version asserts that every semicycle of (α, γ) separates if and only if $g(\alpha, \gamma) = 0$. We now prove that this version follows from ours.

We must show that every semicycle of (α, γ) separates if and only if every normal semicycle of $K(\alpha, \gamma)$ separates. Suppose that every normal semicycle of $K(\alpha, \gamma)$ separates. It is then immediate that every semicycle of (α, γ) separates. Conversely, suppose that every semicycle of (α, γ) separates. Let C' be a normal semicycle of $K(\alpha, \gamma)$, and let D' be a semipath linking its sides. Since C' is normal, the red edges adjacent to a blue edge of C' are either both in $I(C')$ or both in $O(C')$. If these red edges belong to a red-blue square, it follows that no vertex of this square can be a terminal vertex of D'. Therefore, in (α, γ) the semipath corresponding to D' links the sides of the semicycle corresponding to C'. This contradiction establishes that every normal semicycle of $K(\alpha, \gamma)$ separates.

5.3 A condition for a b-cycle to separate

This section is devoted to a proof of Theorem 5.2. The proof consists of several lemmas. We may assume that K is connected since the general case follows by applying the theorem to each component separately.

Lemma 5.5 *Let $C = \sum \mathcal{U}$, where $\mathcal{U}$ is a set of red-blue and blue-yellow bigons in a 3-graph K. Let b be a blue edge, not in $\beta(C)$, joining vertices v and w. Then either $\{v, w\} \subseteq VC$ or $\{v, w\} \cap VC = \emptyset$.*

Proof. Suppose that $v \in VC$. Let a_1 and c_1 be the red and yellow edges, respectively, incident on v. Let a_2 and c_2 be the red and yellow edges, respectively, incident on w. Since $a_1 \in C$, we have $b \in \bigcup \mathcal{U}$. Hence $a_2, c_2 \in C$ and therefore $w \in VC$. Similarly, if $w \in VC$, then $v \in VC$. ■

Lemma 5.6 *If a b-cycle D separates, then D induces a normal b-cycle D' such that no semipath links the sides of D'.*

Proof. Since D separates, D induces a b-cycle $\sum \mathcal{U}$, where $\mathcal{U}$ is a set of bigons. Let $\mathcal{U}_1$ and $\mathcal{U}_2$ denote the set of red-blue and blue-yellow bigons, respectively, included in $\mathcal{U}$, and let $C = \sum(\mathcal{U}_1 \cup \mathcal{U}_2)$. Now consider the cycle $D' = C + \bigcup \mathcal{B}$, where $\mathcal{B}$ is the set of red-yellow bigons included in C. (We include $\bigcup \mathcal{B}$ in the above sum of bigons to ensure that all circuits in D' contain a blue edge.) Since $\emptyset \neq \beta(\sum \mathcal{U}) = \beta(D')$, D' is a b-cycle induced by D. Also, the two edges of D' adjacent to a given blue edge of D' must belong to the same bigon, and hence D' is a normal b-cycle. We claim that no semipath links the sides of D'. Assume by way of contradiction that P is a semipath that links the sides of D'. Let v denote the terminal vertex of P that is in $VI(D')$, and let b be the blue edge incident on v. Thus $b \in P$. Let b join v to w, and let B denote the red-yellow bigon that contains w. By Lemma 5.5, $w \in VC$ and therefore $B \in N(C)$. If $B \in N(D')$, then $P = \{b\}$ and $v, w \in VI(D')$, a contradiction. Hence, we conclude that $B \in \mathcal{B}$ and $B \subseteq C$. Therefore, the blue edge of $P - \{b\}$ incident on a vertex of VB must have both end vertices in VC. Proceeding inductively along P, we find that the terminal vertex x of P other than v must lie in VC. Since the red-yellow bigon that contains x must be in $N(D')$, it follows that $x \in VI(D')$, a contradiction. The lemma follows. ■

Lemma 5.6 proves half of Theorem 5.2. Accordingly, we henceforth assume that D is a b-cycle such that no b-cycle induced by D is a sum of bigons. Let D' be an arbitrary normal b-cycle induced by D. We shall show that the sides of D' are linked by a semipath.

Let $K^\dagger$ be a graph whose vertices are the red-blue and blue-yellow bigons of K and whose edges are the blue edges of K. Any edge $b \in EK^\dagger$ is to join the two bigons containing b in K. Clearly, $K^\dagger$ is connected since K is connected.

Lemma 5.7 *The graph $K^\dagger - \beta(D')$ is connected.*

Proof. Suppose that $K^\dagger - \beta(D')$ is disconnected, and let $L^\dagger$ be a component of $K^\dagger - \beta(D')$. Let us consider the b-cycle $D'' = \sum VL^\dagger + \bigcup \mathcal{B}$, where $\mathcal{B}$ is the set of red-yellow bigons included in $\sum VL^\dagger$. (Recall that $\sum VL^\dagger$ is the symmetric difference of the bigons that constitute the vertex set of $L^\dagger$.) By the construction, $\beta(D'') \subseteq \beta(D') \subseteq \beta(D)$ and therefore D'' is a b-cycle induced by D that is a sum of bigons. This contradicts our assumption. The lemma follows. ■

If there is an edge b of $\beta(K) - \beta(D')$ with an end in $VI(D')$ and one in $VO(D')$, then $\{b\}$ links the sides of D'. Henceforth, we suppose that there is no blue edge with this property, and partition the set $\beta(K) - \beta(D')$ into three classes: the set I of edges with an end in $VI(D')$, the set O of edges with an end in $VO(D')$, and the set M of edges incident with no vertex in $VI(D') \cup VO(D')$.

Lemma 5.8 *The edge sets I and O are non-empty.*

Proof. Assume that I is empty. Then there is no vertex of VD' incident on a red edge and a yellow edge of D'. Since D' is normal, it must therefore be the sum of a set of red-blue and blue-yellow bigons, a contradiction. A similar argument applied to the conjugate of D' shows that $O \neq \emptyset$. ■

Lemma 5.9 *Let P be a path with terminal vertices v and w and blue terminal edges. Suppose that the red-yellow bigons containing v and w are not in $N(P)$.*

Then there exists a semipath P', joining v and w, such that $\beta(P') \subseteq \beta(P)$ and $N(P') \subseteq N(P)$.

Proof. We use induction on $|\beta(P)|$. If $|\beta(P)| = 1$, then P is the required semipath. Now suppose that the lemma holds for all paths with fewer than $|\beta(P)|$ blue edges, where $|\beta(P)| > 1$. Let b denote the blue terminal edge of P incident on v. Let B denote the red-yellow bigon in $N(P)$ containing a vertex x incident on b. (B exists since $|\beta(P)| > 1$.) Let y be the vertex of $VP \cap VB$ that minimises $|P[w, y]|$. Thus $P[w, y]$ is a path Q with fewer blue edges than P. Furthermore, $B \notin N(Q)$ by the choice of y. By the inductive hypothesis, there exists a semipath P' with terminal vertices y and w such that $\beta(P') \subseteq \beta(Q) \subseteq \beta(P)$ and $N(P') \subseteq N(Q) \subseteq N(P)$. Let Q' be a path included in B that joins x and y. Then $P' \cup Q' \cup \{b\}$ is the required semipath. ■

Lemma 5.10 *There exists a semipath that links the sides of D'.*

Proof. Case 1. Suppose that there exists a vertex Y in $VK^\dagger$ incident on an edge $b \in O$ and an edge $b' \in I$. We may choose b and b' so that they are terminal edges of a path P included in the bigon Y such that $\beta(P) - \{b, b'\} \subseteq M \cup D'$. Let $\beta(P) = \{b_1, b_2, \ldots, b_n\}$, where $b_1 = b$ and $b_j \in P[b_i, b_k]$ whenever $i < j < k$. Thus $b_n = b'$. For each $i \in J_{n-1}$ let a_i be the edge of P joining an end w_i of b_i to an end v_{i+1} of b_{i+1}, and let d_i and c_{i+1} be the edges of $EK - Y$ incident on w_i and v_{i+1}, respectively. (See Figure 5.7.)

Suppose that $b_j \in D'$ for some $j < n$. Choose j to be as small as possible subject to this requirement. We have $j > 1$ since $b_1 \notin D'$. Hence $b_{j-1} \in M \cup O$. It follows that $w_{j-1} \notin VD'$, and so $c_j \in D'$. Therefore $d_j \in D'$ since D' is normal. If $j + 1 < n$, then $b_{j+1} \notin M$; hence $\{c_{j+1}, b_{j+1}, d_{j+1}\} \subseteq D'$. By induction $\{c_{n-1}, b_{n-1}, d_{n-1}\} \subseteq D'$. Therefore $a_{n-1} \notin D'$. Since $b' \notin D'$, we obtain the contradiction that $b' \in O$. Hence $\{b_2, b_3, \ldots, b_{n-1}\} \subseteq M$, and so $N(P) \cap N(D') = \emptyset$. By Lemma 5.9, there exists a semipath P' with terminal edges b_1 and b_n such that P' misses D'. Therefore P' links the sides of D', as required.

Case 2. Suppose that there is no bigon in $VK^\dagger$ incident on an edge in I and on an edge in O. By Lemma 5.7 and Lemma 5.8, there exists a path $P^\dagger \subseteq M$ in $K^\dagger - \beta(D')$ with terminal vertices Y_1 and Y_2 such that $Y_1 \cap O \neq \emptyset$ and $Y_2 \cap I \neq \emptyset$. We may assume $P^\dagger$ chosen so that $\beta(L) \cap (I \cup O) = \emptyset$, where $L = \bigcup(VP^\dagger - \{Y_1, Y_2\})$. (Recall that $VP^\dagger$ is the set of bigons in K that make up the vertices in $P^\dagger$.) This choice guarantees that $N(Z) \cap N(D') = \emptyset$ for each

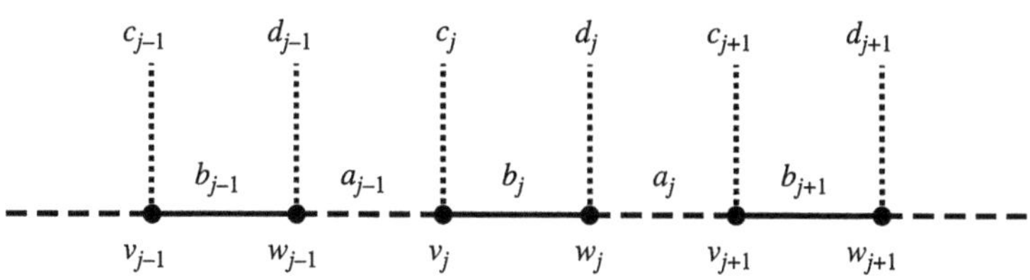

FIGURE 5.7.

internal vertex Z of $P^{\dagger}$. Let P_1 be a path in Y_1 with blue terminal edges $b \in O$ and $b_1 \in (L \cup Y_2) \cap Y_1$ such that the blue internal edges of P_1 are edges in M. Clearly $N(P_1) \cap N(D') = \emptyset$, for otherwise an internal blue edge of P_1 would not be in M. Similarly, let P_2 be a path in Y_2 with blue terminal edges $b' \in I$ and $b_2 \in (L \cup Y_1) \cap Y_2$ such that the blue internal edges of P_2 are edges in M. Again, $N(P_2) \cap N(D') = \emptyset$. It follows that there is a path P in K, with terminal edges b and b', such that $N(P) \cap N(D') = \emptyset$. By Lemma 5.9, we can construct a semipath P' with terminal edges b and b' that misses D'. P' links the sides of D', as required. ■

Lemma 5.10 completes the proof of Theorem 5.2.

5.4 Fundamental sets of semicycles

A set of m b-independent semicycles that does not separate is said to be an *m-fundamental*. With this definition, Theorem 5.1 can be restated in the following way.

Theorem 5.11 *If K is a 3-graph, then the maximum size of an m-fundamental set is $2c(K) - \chi(K)$.*

This section is devoted to a proof of Theorem 5.1. We shall show that the size of a maximum m-fundamental set is

$$2c(K) - \chi(K) = 2c(K) - r(K) + \frac{|VK|}{2}.$$

If $L_1, L_2, \ldots, L_{c(K)}$ denote the components of K, then

$$\begin{aligned} \sum_{i=1}^{c(K)} (2c(L_i) - \chi(L_i)) &= \sum_{i=1}^{c(K)} \left(2 - r(L_i) + \frac{|VL_i|}{2}\right) \\ &= 2c(K) - \chi(K). \end{aligned}$$

Therefore, if Theorem 5.1 holds for each L_i, then it holds for K. We henceforth assume K to be connected, so that $c(K) = 1$.

Lemma 5.12 *If $\mathcal{C}$ is an m-fundamental set, then $m \leq 2 - \chi(K)$.*

Proof. Let C be a semicycle in $\mathcal{C}$ and suppose that a is a red or yellow edge contained in C. Let B denote the red-yellow bigon that contains a. Clearly, $C' = C + B$ is a semicycle such that $a \notin C'$ and $\beta(C') = \beta(C)$. Let $\mathcal{C}' = (\mathcal{C} - \{C\}) \cup \{C'\}$. Since C is not induced by $\mathcal{C} - \{C\}$, then C' is not induced by $\mathcal{C}' - \{C'\}$, and therefore $\mathcal{C}'$ is a b-independent set. Furthermore, $\beta(\bigcup \mathcal{C}) = \beta(\bigcup \mathcal{C}')$, and so $\mathcal{C}'$ is an m-fundamental set such that the number of semicycles in $\mathcal{C}'$ containing a is one less than in $\mathcal{C}$. Proceeding inductively, we obtain an m-fundamental set $\mathcal{C}''$ such that there exists an edge in each red-yellow bigon of K that is not in $\bigcup \mathcal{C}''$.

Since $\mathcal{C}''$ does not separate, each semicycle in $\mathcal{C}''$ is a cycle that does not lie in the boundary space of K. Also, $\mathcal{C}''$ is a linearly independent set of cycles since it is b-independent. Let $\mathcal{F}$ be a set of $r(K) - 1$ bigons in K. By the remark following Lemma 1.35, $\mathcal{F}$ is a basis for the boundary space of K. If $m > 2 - \chi(K)$, then $\mathcal{F} \cup \mathcal{C}''$ is linearly dependent since

$$\begin{aligned} |\mathcal{F} \cup \mathcal{C}''| &= |\mathcal{F}| + |\mathcal{C}''| \\ &> r(K) - 1 + 2 - r(K) + \frac{|VK|}{2} \\ &= |EK| - |VK| + 1 \\ &= \dim \mathcal{C}(K), \end{aligned}$$

and hence there exists a cycle D, belonging to the boundary space, which is a sum of semicycles in $\mathcal{C}''$. By the construction of $\mathcal{C}''$, it is impossible for D to include a red-yellow bigon. We conclude that D must be a b-cycle. Moreover, D is induced by $\mathcal{C}''$, which is a contradiction. Hence, we conclude that $m \leq 2 - \chi(K)$. ■

The rest of this section is concerned with the construction of a $(2 - \chi(K))$-fundamental set $\mathcal{C}$. This construction together with Lemma 5.12 gives us our theorem. It also implies that $\mathcal{F} \cup \mathcal{C}$ is a basis for the cycle space of K, where $\mathcal{F}$ is a set of $r(K) - 1$ bigons in K.

Let K be a 3-graph with a blue 1-dipole b. Let C' be a semicycle in the 3-graph $K' = K - [b]$. The following uses the notation in Figure 3.1 (p. 40). If C' does not meet A', then all of the edges of C' are in K and we define $C = C'$. If C' meets A', then let x and y be the two poles of A' with respect to C'. Let $P' = A - \{a_1, c_1\}$ and $Q' = B - \{a_2, c_2\}$. Assume without loss of generality that $x \in VP'$ and consider the following two cases.

Case 1. $y \in VP'$. Let P be a path in A that joins x to y. Then, we define $C = (C' - A') \cup P$.

Case 2. $y \in VQ'$. Let P be a path in A that joins x to v and Q a path in B that joins w to y. We define $C = (C' - A') \cup P \cup Q \cup \{b\}$.

In the cases presented above, C is clearly a semicycle in K such that $\beta(C) - \{b\} = \beta(C')$. We say that C is a semicycle (in K) *implied* by the semicycle C' (in K').

More generally, let $\mathcal{C}' = \{C'_1, C'_2, \ldots, C'_n\}$ be a set of semicycles in K'. For each i let C_i be a semicycle in K implied by C'_i, and let $\mathcal{C} = \{C_1, C_2, \ldots, C_n\}$. We say that $\mathcal{C}$ is a set of semicycles (in K) *implied* by the set $\mathcal{C}'$ of semicycles (in K'). Clearly, $\beta(\bigcup \mathcal{C}) - \{b\} = \beta(\bigcup \mathcal{C}')$.

Lemma 5.13 *Let K be a 3-graph with blue 1-dipole b, and let $K' = K - [b]$. If $\mathcal{C}'$ is an m-fundamental set in K', then a set $\mathcal{C}$ of semicycles in K implied by $\mathcal{C}'$ is an m-fundamental set in K.*

Proof. First, we show that $\mathcal{C}$ is b-independent. Suppose not, and let C be a semicycle in $\mathcal{C}$ that is induced by $\mathcal{C} - \{C\}$. Therefore, $\beta(C) \subseteq \beta(\bigcup(\mathcal{C} - \{C\}))$. Let C' denote the semicycle in $\mathcal{C}'$ that implies C. The fact that $\beta(C') = \beta(C) - \{b\} \subseteq \beta(\bigcup(\mathcal{C} - \{C\})) - \{b\} = \beta(\bigcup(\mathcal{C}' - \{C'\}))$ implies that $\mathcal{C}'$ is not b-independent, a contradiction. Hence, we conclude that $\mathcal{C}$ is b-independent.

The following uses the notation of Figure 3.1. Let Y denote the red-blue bigon in K that includes $\{a_1, b, a_2\}$, and let Y' be the red-blue bigon $(Y - \{a_1, b, a_2\}) \cup \{a\}$ in K'. Similarly, let R denote the blue-yellow bigon in K that includes $\{c_1, b, c_2\}$, and let R' be the blue-yellow bigon $(R - \{c_1, b, c_2\}) \cup \{c\}$ in K'.

Suppose that $\mathcal{C}$ induces a b-cycle $D = \sum \mathcal{U}$, where $\mathcal{U}$ is a set of bigons. Let $\mathcal{U}_1$ and $\mathcal{U}_2$ denote the set of red-blue and blue-yellow bigons, respectively, included in $\mathcal{U}$. Let $D_1 = \sum(\mathcal{U}_1 \cup \mathcal{U}_2) + \bigcup \mathcal{B}$, where $\mathcal{B}$ is the set of red-yellow bigons included in $\sum(\mathcal{U}_1 \cup \mathcal{U}_2)$. Then D_1 is also a b-cycle induced by $\mathcal{C}$ that separates. If $Y \in \mathcal{U}_1$, then let $\mathcal{U}_1' = (\mathcal{U}_1 - \{Y\}) \cup \{Y'\}$; otherwise, let $\mathcal{U}_1' = \mathcal{U}_1$. If $R \in \mathcal{U}_2$, then let $\mathcal{U}_2' = (\mathcal{U}_2 - \{R\}) \cup \{R'\}$; otherwise, let $\mathcal{U}_2' = \mathcal{U}_2$. Let $D' = \sum(\mathcal{U}_1' \cup \mathcal{U}_2') + \bigcup \mathcal{B}'$, where $\mathcal{B}'$ is the set of red-yellow bigons included in $\sum(\mathcal{U}_1' \cup \mathcal{U}_2')$. If $D' = \emptyset$, then D would have exactly one blue edge, namely b. By the definition of a b-cycle, D would consist of one circuit comprising b and some red and yellow edges that belong to a red-yellow bigon, in contradiction to the fact that b is a blue 1-dipole. Hence we conclude that D' is a b-cycle that separates K'. Moreover, D' is induced by $\mathcal{C}'$ since $\beta(D') = \beta(D_1) - \{b\}$. This result is a contradiction since $\mathcal{C}'$ does not separate. Hence, we conclude that $\mathcal{C}$ is an m-fundamental set in K. ■

Now suppose that K' is obtained from K by a finite sequence of blue 1-dipole cancellations, and that C' is a semicycle in K'. Then we apply the definition of an implied semicycle inductively to obtain a semicycle C in K that is *implied* by C'. Similarly, we speak of a set of semicycles in K' *implying* a set of semicycles in K. By Lemma 5.13, if $\mathcal{C}'$ is an m-fundamental set in K', then a set $\mathcal{C}$ of semicycles in K implied by $\mathcal{C}'$ is an m-fundamental set in K.

Suppose K to be a 3-graph with a unique red-yellow bigon B. Let b be a blue edge in K joining vertices x and y. Let P be a path in B joining x and y. The blue edge b can be used to define a semicycle in K, namely $C = \{b\} \cup P$. We say that C is a semicycle *formed* from b.

Recall that $K^\dagger$ is a graph whose vertices are the red-blue and blue-yellow bigons of K and whose edges are the blue edges of K. Any edge $b \in EK^\dagger$ joins the two bigons containing b in K. Let T be a spanning tree of $K^\dagger$, and let $T' = EK^\dagger - ET$.

Lemma 5.14 $|T'| = 2 - \chi(K)$.

Proof. Since K has exactly one red-yellow bigon, we have

$$|VK^\dagger| = r(K) - 1.$$

The number of edges in a spanning tree for the graph $K^\dagger$ is $|VK^\dagger| - 1 = r(K) - 2$. Observe that $|EK^\dagger| = |\beta(K)| = |VK|/2$. Hence

$$\begin{aligned} |T'| &= |EK^\dagger| - |ET| \\ &= \frac{|VK|}{2} - (r(K) - 2) \\ &= 2 - \chi(K). \; ■ \end{aligned}$$

Lemma 5.15 *If K is a connected 3-graph with one red-yellow bigon, then there exists a $(2 - \chi(K))$-fundamental set in K.*

Proof. Let $j = 2 - \chi(K) = |T'|$, where T' is defined as above. For each $b_i \in T'$, where $i \in J_j$, let C_i be a semicycle formed from b_i. Let $\mathcal{C} = \{C_1, C_2, \ldots, C_j\}$. Hence $\beta(\bigcup \mathcal{C}) = T'$. Since all of the b_i's are distinct, $\mathcal{C}$ is a set of j b-independent semicycles in K. Assume that $\mathcal{C}$ induces a b-cycle that separates. Then $\mathcal{C}$ induces a b-cycle D of the form $\sum \mathcal{U}$ for some set $\mathcal{U}$ of bigons in K. Since we may add the red-yellow bigon to $\sum \mathcal{U}$ and still have a b-cycle induced by $\mathcal{C}$ that separates, we may assume that $\mathcal{U}$ does not contain the red-yellow bigon. Therefore $\beta(D) = \partial_{K^\dagger}\mathcal{U}$. Hence, there must exist an edge in T that is in $\beta(D)$. However, this is impossible since $\beta(D) \subseteq \beta(\bigcup \mathcal{C}) = T'$. Therefore $\mathcal{C}$ is a $(2 - \chi(K))$-fundamental set in K. ■

We now combine the results of the preceding lemmas to produce a proof of Theorem 5.1.

Proof. By Lemma 5.12, we have $m \leq 2 - \chi(K)$. Hence, we are required to show that there exists a $(2 - \chi(K))$-fundamental set in K. Cancel blue 1-dipoles from K one at a time until none is left, and let K' denote the resulting graph. Therefore K' has exactly one red-yellow bigon. By Lemma 5.15, there exists a $(2 - \chi(K'))$-fundamental set $\mathcal{C}'$ in K'. Since K' is obtained from K by a finite sequence of blue 1-dipole cancellations, by Lemma 5.13 a set of semicycles in K implied by $\mathcal{C}'$ is a $(2 - \chi(K))$-fundamental set in K. ■

6

Mac Lane's Theorem for 3-Graphs

In this chapter, we give our first characterisation of planar graphs. This characterisation is due to Mac Lane. The basic idea is that a 2-connected graph is planar if and only if there is a family of circuits enjoying properties that we intuitively expect the boundaries of faces to have. The work in this chapter is in the more general setting of 3-graphs. Mac Lane's theorem is deduced as a corollary. We conclude the chapter by obtaining another criterion due to Whitney.

6.1 Congruence

Recall that $\beta(K)$ denotes the set of blue edges of a 3-graph K. Similarly, we denote by $\rho(K)$ the set of red edges of K.

A basic tool in this chapter is the notion of congruence of 3-graphs. Let K and L be two 3-graphs. Suppose that there exist bijections θ, φ, and σ between $\mathcal{B}(L)$ and $\mathcal{B}(K)$, $\beta(L)$ and $\beta(K)$, and $\rho(L)$ and $\rho(K)$, respectively. Furthermore, suppose that

(1) for any red-yellow bigon B in L and any red edge $a \in B$ we have $\sigma(a) \in \theta(B)$, and

(2) for any blue edge b in L adjacent to a red edge a we have $\varphi(b)$ adjacent to $\sigma(a)$.

Then K and L are *congruent*. Thus, two gems are congruent if and only if some graph underlies both of them. Moreover, if K and L are congruent, then by condition (2) we have a bijection between the red-blue bigons of K and the red-blue bigons of L. Evidently, congruence is an equivalence relation.

Example 6.1 *Consider the gems L and K in Figure 6.1. Let* B_1, B_2, B_3, *and* B_4 *denote the red-yellow bigons* $\{a_1, c_1, a_2, c_2\}$, $\{a_3, c_3, a_4, c_4\}$, $\{a_5, c_5, a_6, c_6\}$, *and* $\{a_7, c_7, a_8, c_8\}$, *respectively. Define* $\theta(B_i) = B_{i+2}$ *for each* $i \in J_2$, $\varphi(b_i) = b_{i+4}$ *for each* $i \in J_4$, *and* $\sigma(a_i) = a_{i+4}$ *for each* $i \in J_4$. *Now, observe that*

(1) *for any red-yellow bigon* B_j *in L and any red edge* $a_i \in B_j$ *we have* $\sigma(a_i) = a_{i+4} \in \theta(B_j) = B_{j+2}$, *and*

(2) *for any blue edge* b_i *in L adjacent to a red edge* a_j *we have* $\varphi(b_i) = b_{i+4}$ *adjacent to* $\sigma(a_j) = a_{i+4}$.

Hence, we conclude that K and L are congruent. As expected, a graph G underlies both, and this is the connected graph with two vertices, two links, and no loops. L models an imbedding of G in the sphere and K models an imbedding of G in the projective plane (the non-orientable surface of genus 1).

If K and L are congruent 3-graphs and $\{e_1, e_2, \ldots, e_n\}$ is a set S of red (blue) edges in L, then for conciseness we usually write S for $\sigma[S]$ ($\varphi[S]$) and e_i for $\sigma(e_i)$ ($\varphi(e_i)$) when no ambiguity results.

6.2 Semicycle covers

A *red-refined* 3-graph is one in which all blue-yellow bigons are semicycles. A set $\mathcal{S}$ of semicycles in a 3-graph K is said to be a *semicycle cover* if the set of intersections of members of $\mathcal{S}$ with $\beta(K)$ is a partition of $\beta(K)$. In other words, each blue edge of K belongs to a unique member of $\mathcal{S}$. Hence $\mathcal{R}(K)$ is a semicycle

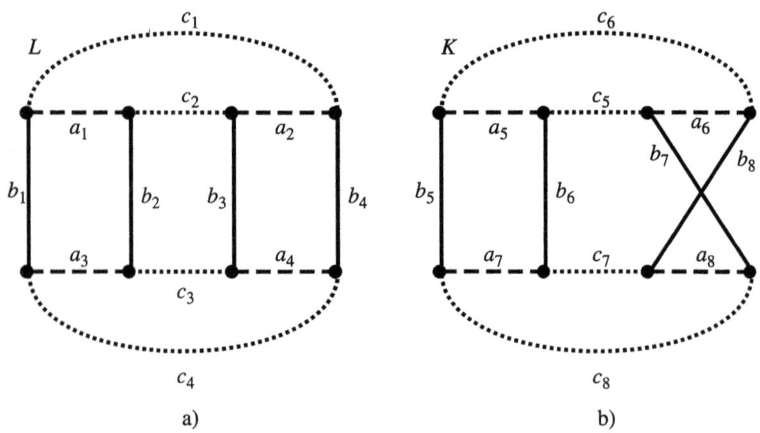

FIGURE 6.1.

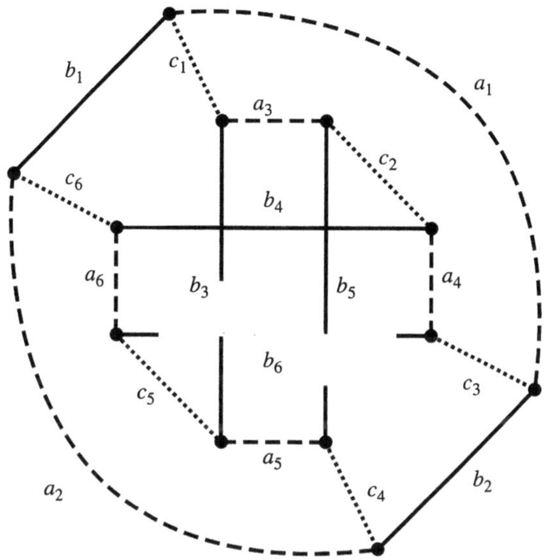

FIGURE 6.2.

cover if K is red-refined. If $\mathcal{B}(K) \cup \mathcal{Y}(K) \cup \mathcal{S}$ spans the cycle space of K, we say that $\mathcal{S}$ is a *spanning* semicycle cover.

Example 6.2 *Consider the gem K of Figure 6.2. Let $C_1 = \{b_1, c_1, a_3, c_2, b_4, c_6\}$, $C_2 = \{b_3, a_3, c_2, a_4, b_6, c_5\}$, and $C_3 = \{b_5, c_2, a_4, c_3, b_2, c_4\}$. One can easily check that $\mathcal{B}(K) \cup \mathcal{Y}(K) \cup \{C_1, C_2, C_3\}$ spans the cycle space of K. Therefore $\{C_1, C_2, C_3\}$ is a spanning semicycle cover.*

In this chapter, we prove that a 3-graph K is congruent to a spherical 3-graph if and only if there exists a spanning semicycle cover in K. We employ the operations of cancellation and creation of squares as described in Section 3.1 (p. 42) but with the colours red and yellow interchanged. Thus, the squares cancelled or created are blue-yellow ones.

Let K be a 3-graph. Suppose that there is a blue-yellow bigon R in K that is not a semicycle. Then there is a red-yellow bigon B such that $B \cap R$ contains two yellow edges c and c'. Hence, $B - \{c, c'\}$ is the union of two disjoint paths P_1 and P_2, each of which contains a red edge. Let a_1 be a red edge in P_1 and let a_2 be a red edge in P_2. Split a_1 and a_2 to create a blue-yellow bigon R' and let K' denote the resulting graph. The red-yellow bigon B now has been replaced by two red-yellow bigons A and A' in K'. Furthermore, R is a blue-yellow bigon in K' that meets A in fewer edges than R meets B. A similar statement holds for A'. We also note that R' is a semicycle in K' and that any blue-yellow bigons in K' that are semicycles in K are semicycles in K'. Proceeding inductively, we obtain a 3-graph such that any blue-yellow bigon meets each red-yellow bigon in, at most, one edge and hence is a semicycle. We conclude that it is possible to construct a 3-graph L from K by blue 1-dipole and blue-yellow 2-dipole creations such that

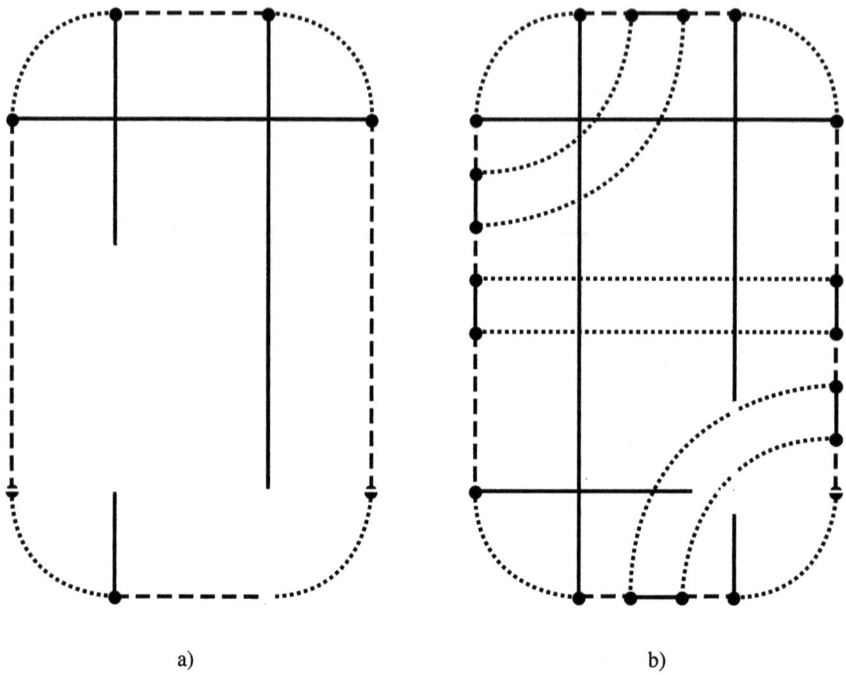

FIGURE 6.3.

all blue-yellow bigons in L are semicycles. Hence L is a red-refined 3-graph. We say that L is a *red-refinement* of K.

Example 6.3 *Consider the gem K in Figure 6.3a. Figure 6.3b illustrates a red-refinement of K.*

6.3 Boundary covers

In this section, we prove our main theorem in one direction. That is, any 3-graph congruent to a spherical 3-graph has a spanning semicycle cover. We prepare for its proof with some preliminary definitions and lemmas. A semicycle cover $\mathcal{S}$ is a *boundary cover* if $\mathcal{B}(K) \cup \mathcal{Y}(K) \cup \mathcal{S}$ spans the boundary space of K. By Theorem 5.3 (p. 69), boundary covers are also spanning semicycle covers in spherical 3-graphs.

Lemma 6.1 *Let C and D be two cycles in a 3-graph K such that $\beta(C) = \beta(D)$. Then $C = D + \bigcup \mathcal{B}$ for some set $\mathcal{B}$ of red-yellow bigons.*

Proof. Since $\beta(C) = \beta(D)$, $C + D$ has no blue edges and is therefore a union of red-yellow bigons. The result follows immediately. ■

Lemma 6.2 *Let K be a 3-graph with blue 1-dipole b and let $L = K - [b]$. There exists a boundary cover in L if there exists a boundary cover in K.*

Proof. The following uses the notation of Figure 3.1 (p. 40). Let $\mathcal{S}$ denote a boundary cover of K. Suppose that there exists a semicycle $C \in \mathcal{S}$ which contains c_1. Then clearly $C + A$ is a semicycle C_1 such that $c_1 \notin C_1$ and $c_2 \in C_1$ if and only if $c_2 \in C$. Furthermore, $\mathcal{S}_1 = (\mathcal{S} - \{C\}) \cup \{C_1\}$ is a boundary cover of K. A similar argument may be applied to c_2. Therefore, we assume that no semicycle in $\mathcal{S}$ contains c_1 or c_2. In particular, some semicycle C_2 in $\mathcal{S}$ includes $\{a_1, b, a_2\}$.

Let $\mathcal{S}_2 = (\mathcal{S} - \{C_2\}) \cup \{C_3\}$, where $C_3 = (C_2 - \{a_1, b, a_2\}) \cup \{a\}$. Evidently, the set of intersections of members of $\mathcal{S}_2$ with $\beta(L)$ is a partition of $\beta(L)$. We claim that $\mathcal{B}(L) \cup \mathcal{Y}(L) \cup \mathcal{S}_2$ spans the boundary space of L. It is sufficient to show that any blue-yellow bigon R of L is a sum of circuits in $\mathcal{B}(L) \cup \mathcal{Y}(L) \cup \mathcal{S}_2$. If $c \in R$, let $R_1 = (R - \{c\}) \cup \{c_1, b, c_2\}$; otherwise, let $R_1 = R$. Then R_1 is a blue-yellow bigon in K, and hence $R_1 = \sum \mathcal{U}$ for some set $\mathcal{U}$ of circuits in $\mathcal{B}(K) \cup \mathcal{Y}(K) \cup \mathcal{S}$. We claim that $A \in \mathcal{U}$ if and only if $B \in \mathcal{U}$. However, this is clear, for $c_2 \in R_1$ if and only if $c_1 \in R_1$ and no semicycle in $\mathcal{S}$ contains either c_1 or c_2. Let Y denote the red-blue bigon in K that includes $\{a_1, b, a_2\}$, and let Y' denote the red-blue bigon in L that contains a. Let $\mathcal{U}_1$ be the set obtained from $\mathcal{U}$ by replacing C_2 with C_3, A and B with A', and Y with Y' if necessary. Evidently, $R = \sum \mathcal{U}_1$ and $\mathcal{U}_1 \subseteq \mathcal{B}(L) \cup \mathcal{Y}(L) \cup \mathcal{S}_2$, as required.

If $\mathcal{S}_2$ is a set of semicycles, then we are done. Clearly, C_3 is a semicycle in L. Suppose that there is a circuit D in $\mathcal{S}_2$ that is not a semicycle in L. Then A' has just four poles with respect to D, for otherwise D is not a semicycle in K. Let v, w, x, y denote the four poles, where $A'_y[v, w] \subset D$ and $x \in A'_v[w, y]$. Thus $A'_w[x, y] \subseteq D$. Clearly, $C_4 = D_w[v, y] \cup A'_w[v, y]$ and $C_5 = D_v[w, x] \cup A'_v[w, x]$ are semicycles in L. Since $\beta(D)$ is the disjoint union of $\beta(C_4)$ and $\beta(C_5)$, then by Lemma 6.1, $C_4 + C_5 = D + \cup\mathcal{B}$ for some set $\mathcal{B}$ of red-yellow bigons in L. We conclude that $\mathcal{S}_3 = (\mathcal{S}_2 - \{D\}) \cup \{C_4, C_5\}$ is a set of circuits, with more semicycles than $\mathcal{S}_2$, such that $\mathcal{B}(L) \cup \mathcal{Y}(L) \cup \mathcal{S}_3$ spans the boundary space of L. Moreover, the set of intersections of members of $\mathcal{S}_3$ with $\beta(L)$ is a partition of $\beta(L)$. Proceeding inductively ,we obtain a boundary cover of L. ■

Lemma 6.3 *Let K be a 3-graph with blue-yellow 2-dipole $\{a, b\}$ and let $L = K - [a, b]$. Then there exists a boundary cover in L if there exists a boundary cover in K.*

Proof. The following uses the notation of Figure 3.2 (p. 41) but with the colours red and yellow interchanged. Let $\mathcal{S}$ denote a boundary cover of K. Let B be the red-yellow bigon in K that includes $\{c_1, a, c_2\}$. Let C denote the semicycle in $\mathcal{S}$ that contains b. Suppose that $C \neq \{a, b\}$. Then clearly $\{a, b\} = C + B$. Furthermore, $\mathcal{S}_1 = (\mathcal{S} - \{C\}) \cup \{C + B\}$ is a boundary cover of K. Therefore, we assume that the semicycle C in $\mathcal{S}$ that contains b is $\{a, b\}$.

Let $\mathcal{S}' = \mathcal{S} - \{C\}$. Suppose that there exists a semicycle $C_1 \in \mathcal{S}'$ which contains a. Then clearly $C_2 = C_1 + B$ is a semicycle such that $a \notin C_2$. Furthermore, $\mathcal{S}_1 = (\mathcal{S} - \{C_1\}) \cup \{C_2\}$ is a boundary cover of K. Therefore, we assume that no semicycle in $\mathcal{S}'$ contains a. Hence $\mathcal{S}'$ is a set of semicycles in L that do not contain c.

Evidently, the set of intersections of members of $\mathcal{S}'$ with $\beta(L)$ is a partition of $\beta(L)$. We claim that $\mathcal{B}(L) \cup \mathcal{Y}(L) \cup \mathcal{S}'$ spans the boundary space of L, and hence that $\mathcal{S}'$ is a boundary cover of L. It is sufficient to show that any blue-yellow bigon R of L is a sum of circuits in $\mathcal{B}(L) \cup \mathcal{Y}(L) \cup \mathcal{S}'$. Clearly, R is a blue-yellow bigon in K, and hence $R = \sum \mathcal{U}$ for some set $\mathcal{U}$ of circuits in $\mathcal{B}(K) \cup \mathcal{Y}(K) \cup \mathcal{S}$. Let Y denote the red-blue bigon in K that contains b. Since no semicycle in $\mathcal{S}'$ contains a or b, we have $a \notin \bigcup(\mathcal{U} - \{B, C\})$ and $b \notin \bigcup(\mathcal{U} - \{Y, C\})$. Therefore, the fact that $R \neq \{a, b\}$ implies that either $\{B, C, Y\} \subseteq \mathcal{U}$ or $\{B, C, Y\} \cap \mathcal{U} = \emptyset$.

Let B' and Y' denote the red-yellow and red-blue bigons, respectively, in L that contain c. If $\{B, C, Y\} \subseteq \mathcal{U}$, then let $\mathcal{U}' = (\mathcal{U} - \{B, C, Y\}) \cup \{B', Y'\}$; otherwise, let $\mathcal{U}' = \mathcal{U}$. Evidently, $R = \sum \mathcal{U}'$ and $\mathcal{U}' \subseteq \mathcal{B}(L) \cup \mathcal{Y}(L) \cup \mathcal{S}'$, as required. ■

Theorem 6.4 *Every 3-graph has a boundary cover.*

Proof. Let K be a red-refinement of an arbitrary 3-graph L. Therefore, $\mathcal{R}(K)$ is a set of semicycles in K. It is immediate that $\mathcal{R}(K)$ is a boundary cover of K since the boundary space is the space spanned by $\mathcal{B}(K) \cup \mathcal{Y}(K) \cup \mathcal{R}(K)$. Hence, our theorem follows from Lemmas 6.2 and 6.3, since L is obtained from K by a finite sequence of blue 1-dipole and blue-yellow 2-dipole cancellations. ■

Let G be the graph obtained from a 3-graph K by contracting the red-yellow bigons to single vertices. In other words, G is the contraction of K determined by the vertex sets of the red-yellow bigons. We say that G is the *red-yellow reduction* of K. If K and L are congruent 3-graphs, then evidently the red-yellow reductions of K and L are the same graph. We say that cycles C in G and D in K *correspond* if $C = \beta(D)$.

Lemma 6.5 *Let G be the red-yellow reduction of a 3-graph K. Let $\mathcal{C}$ be a set of circuits in G and let $\mathcal{D}$ be a corresponding set of semicycles in K. Then $\mathcal{C}$ spans $\mathcal{C}(G)$ if and only if $\mathcal{B}(K) \cup \mathcal{D}$ spans $\mathcal{C}(K)$.*

Proof. First, assume that $\mathcal{C}$ spans $\mathcal{C}(G)$ and let D be a circuit in K. We may assume that D contains a blue edge for otherwise D is a red-yellow bigon. Evidently, $\beta(D)$ is a cycle in G and therefore $\beta(D) = \sum \mathcal{U}$ for some subset $\mathcal{U}$ of $\mathcal{C}$. Let $\mathcal{V}$ be the set of semicycles in $\mathcal{D}$ that correspond to the circuits in $\mathcal{U}$. Then clearly $\beta(\sum \mathcal{V}) = \beta(D)$. By Lemma 6.1, $D = \sum \mathcal{V} + \sum \mathcal{B}$ for some set $\mathcal{B} \subseteq \mathcal{B}(K)$. We conclude that $\mathcal{B}(K) \cup \mathcal{D}$ spans $\mathcal{C}(K)$.

Now, assume that $\mathcal{B}(K) \cup \mathcal{D}$ spans $\mathcal{C}(K)$ and let C be a circuit in G. Let D be a semicycle in K such that $\beta(D) = C$. Then $D = \sum \mathcal{V}$ for some subset $\mathcal{V}$ of $\mathcal{B}(K) \cup \mathcal{D}$. Let $\mathcal{U}$ be the set of circuits in $\mathcal{C}$ that correspond to the semicycles in $\mathcal{V} \cap \mathcal{D}$. Evidently, $C = \beta(D) = \beta(\sum \mathcal{V}) = \beta(\sum(\mathcal{V} \cap \mathcal{D})) = \sum \mathcal{U}$. We conclude that $\mathcal{C}$ spans $\mathcal{C}(G)$, as required. ■

Theorem 6.6 *If K is congruent to a spherical 3-graph, then there exists a spanning semicycle cover in K.*

Proof. Let L be a spherical 3-graph congruent to K, and let G be the red-yellow reduction of K and L. By Theorem 6.4, there exists a boundary cover $\mathcal{S}'$ of L. Since L is spherical, $\mathcal{S}'$ is a spanning semicycle cover of L. Let $\mathcal{Y}$ be the set of cycles in G that correspond to the red-blue bigons in L. Let $\mathcal{R}$ be the set of circuits in G that correspond to the semicycles in $\mathcal{S}'$. By Lemma 6.5, we conclude that $\mathcal{Y} \cup \mathcal{R}$ spans $\mathcal{C}(G)$.

For each circuit $R \in \mathcal{R}$, choose a semicycle in K that corresponds to R, and let $\mathcal{S}$ be the set of all semicycles so chosen. The fact that $\mathcal{R}$ is a partition of EG establishes that the set of intersections of members of $\mathcal{S}$ with $\beta(K)$ is a partition of $\beta(K)$. Let D be a circuit in K. Then $\beta(D)$ is a cycle in G. Therefore, $\beta(D) = \sum \mathcal{U}$ for some $\mathcal{U} \subseteq \mathcal{Y} \cup \mathcal{R}$. Let $\mathcal{V}$ be the set of semicycles in $\mathcal{S}$ that correspond to the cycles in $\mathcal{U} \cap \mathcal{R}$, and let $\mathcal{W}$ be the set of red-blue bigons in K that correspond to the cycles in $\mathcal{U} \cap \mathcal{Y}$. Therefore $\beta(D) = \sum \mathcal{U} = \beta(\sum(\mathcal{V} \cup \mathcal{W}))$. Hence, by Lemma 6.1, $D = \sum(\mathcal{V} \cup \mathcal{W}) + \bigcup \mathcal{B}$, for some set $\mathcal{B}$ of red-yellow bigons in K. Thus D is a sum of bigons in $\mathcal{B}(K) \cup \mathcal{Y}(K) \cup \mathcal{S}$. We conclude that $\mathcal{S}$ is a spanning semicycle cover in K, as required. ■

6.4 Partial congruence

We now generalise the concept of congruence that was given in Section 6.1. Let K and L be two 3-graphs. Suppose that there exist a partition $\mathcal{Q}$ of $\mathcal{B}(L)$ and bijections θ, φ, σ between $\mathcal{Q}$ and $\mathcal{B}(K)$, $\beta(L)$ and $\beta(K)$, and $\rho(L)$ and $\rho(K)$, respectively. Furthermore, suppose that

(1) for any cell $\mathcal{B}$ of $\mathcal{Q}$ and any red edge $a \in \bigcup \mathcal{B}$ we have $\sigma(a) \in \theta(\mathcal{B})$, and

(2) for any blue edge b in L adjacent to a red edge a we have $\varphi(b)$ adjacent to $\sigma(a)$.

Then we say that L is *partially congruent* to K (with respect to the partition $\mathcal{Q}$).

If L is partially congruent to K and $\{e_1, e_2, \ldots, e_n\}$ is a set S of red (blue) edges in L, then for conciseness we usually write S for $\sigma[S](\varphi[S])$ and e_i for $\sigma(e_i)(\varphi(e_i))$ when no ambiguity results.

Let L be partially congruent to a 3-graph K and let $\mathcal{S}$ be a set of cycles in K. If for each $C \in \mathcal{S}$ there exists a cycle D in L such that $\beta(D) = \beta(C)$, then we say that L is *faithful* to K (with respect to $\mathcal{S}$). The cycles D and C are said to *correspond* to each other.

Example 6.4 *Consider the 3-graphs K and L in Figure 6.4. Let* $B_1 = \{a_1, c_3\}$, $B_2 = \{a_4, c_4\}$, *and* $B_3 = \{a_2, a_3, c_1, c_2\}$. *Let* $\mathcal{Q}$ *be the partition* $\{\{B_1, B_2\}, \{B_3\}\}$ *of* $\mathcal{B}(L)$. *From the labelling of the edges of K, it is clear that L is partially congruent to K with respect to the partition* $\mathcal{Q}$. *Let C be the cycle* $\{b_2, c_4, b_3, c_1\}$ *in K. Since* $|\partial_L V B_1 \cap \beta(C)| = 1$, $\beta(C)$ *is not the blue edge set of a cycle in L. Hence L is not faithful to K with respect to* $\{C\}$.

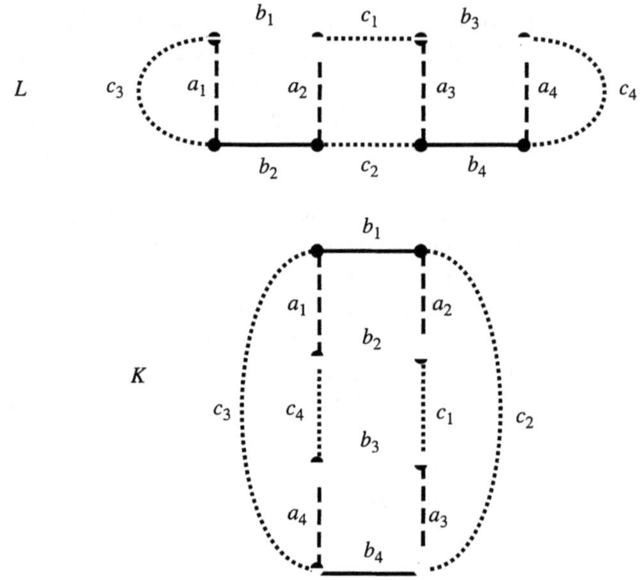

FIGURE 6.4.

Lemma 6.7 *Let L be partially congruent to a 3-graph K, and let C be a cycle in L. Then $\beta(C)$ is the blue edge set of a cycle in K.*

Proof. This follows from the fact that $|\beta(C) \cap \partial VA|$ is even for all $A \in \mathcal{B}(L)$, and therefore $|\beta(C) \cap \partial VB|$ is even for all $B \in \mathcal{B}(K)$. ■

Lemma 6.8 *Let L be a 3-graph partially congruent to a 3-graph K with respect to a partition $\mathcal{Q}$. If L is faithful to K with respect to a spanning semicycle cover $\mathcal{S}$, then no two red-yellow bigons in a cell of $\mathcal{Q}$ belong to the same component of L.*

Proof. We shall prove the contrapositive. Suppose that B_1 and B_2 are distinct red-yellow bigons in a cell $\mathcal{B}$ of $\mathcal{Q}$ that belong to the same component L_1 of L. Since L_1 is connected, there exists a path P, joining a vertex in VB_1 to a vertex in VB_2, such that $|P \cap \partial VB_1| = |P \cup \partial VB_2| = 1$ and $|P \cap \partial VB|$ is even for all $B \in \mathcal{B}(L) - \{B_1, B_2\}$. Therefore $|P \cap \partial VB'|$ is even for all red-yellow bigons B' in K, since B_1 and B_2 both correspond to a single red-yellow bigon. Hence $\beta(P)$ is the blue edge set of a cycle in K.

Since $\mathcal{S}$ is a spanning semicycle cover, $\beta(P)$ is the blue edge set of a cycle $\sum(\mathcal{Y} \cup \mathcal{U})$ for sets $\mathcal{Y} \subseteq \mathcal{Y}(K)$ and $\mathcal{U} \subseteq \mathcal{S}$. Let $\mathcal{R}$ be a set of cycles in L that correspond to the cycles in $\mathcal{U}$. Evidently, $\beta(P) = \beta(\sum(\mathcal{Y} \cup \mathcal{R}))$, and $\sum(\mathcal{Y} \cup \mathcal{R})$ is a cycle in L. Hence $|P \cap \partial VB|$ is even for all $B \in \mathcal{B}(L)$, a contradiction to the fact that $|P \cap \partial VB_1| = |P \cap \partial VB_2| = 1$. ■

Example 6.5 *Consider the* 3*-graphs K and L in Figures 6.5a and b, respectively. Let*

$$\begin{aligned} C_1 &= \{b_5, c_8, a_7, b_8, a_8, c_6, a_6\}, \\ C_2 &= \{b_6, c_7, a_7, b_7, c_6\}, \\ C_3 &= \{b_1, a_2, c_3, a_4, b_4, c_2, a_1\}, \\ C_4 &= \{b_2, c_3, b_3, a_3, c_2\}. \end{aligned}$$

Let $\mathcal{S} = \{C_1, C_2, C_3, C_4\}$. *One can easily check that* $\mathcal{S}$ *is a spanning semicycle cover in* K. *Let* B_1, B_2, B_3, *and* B_4 *be the red-yellow bigons in* L *that contain* a_1, a_2, a_6, *and* a_5, *respectively. Let* $\mathcal{Q}$ *be the partition* $\{\{B_1\}, \{B_2, B_3\}, \{B_4\}\}$ *of* $\mathcal{B}(L)$. *It is clear that L is partially congruent to K with respect to the partition* $\mathcal{Q}$. *Furthermore, L is faithful to K with respect to* $\mathcal{S}$. *We note that* B_2 *and* B_3 *belong to distinct components of L, in agreement with Lemma 6.8.*

Let $\mathcal{S}$ be a semicycle cover in a 3-graph K. The semicycles of $\mathcal{S}$ induce a partition $\mathcal{V}$ of VK into pairs, where two vertices belong to the same cell if and

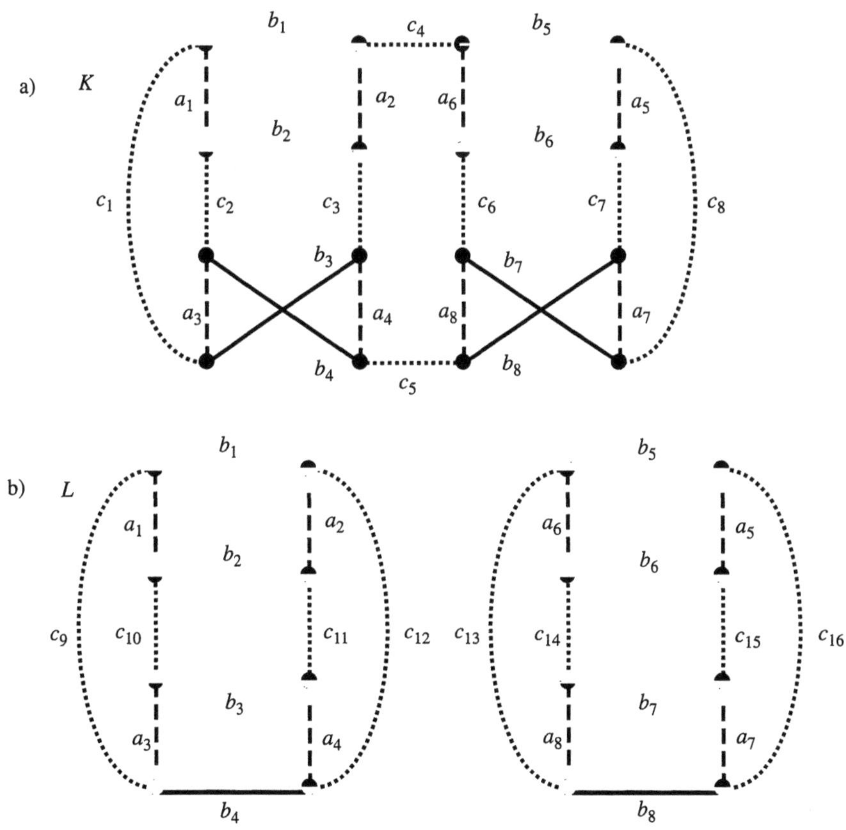

FIGURE 6.5.

only if they are the two poles of a red-yellow bigon with respect to a semicycle in $\mathcal{S}$. Let L be the 3-graph obtained from K by deleting the yellow edges and inserting a yellow edge joining the vertices v and w for each $\{v, w\} \in \mathcal{V}$. We say that L *encodes* K (with respect to $\mathcal{S}$). Clearly L is partially congruent to K with respect to a partition $\mathcal{Q}$. We say that $\mathcal{Q}$ is the *encoding partition*. Furthermore, we have a one to one correspondence between the semicycles in $\mathcal{S}$ and the blue-yellow bigons in L. Hence L is faithful to K with respect to $\mathcal{S}$.

Example 6.6 *Returning to Example 6.5, we see that L is in fact the 3-graph that encodes K with respect to $\mathcal{S}$, and that $\mathcal{Q}$ is the encoding partition. We also note that L is spherical. Lemma 6.9 below states that L will be spherical in general whenever $\mathcal{S}$ is a spanning semicycle cover.*

Lemma 6.9 *Let L be the 3-graph that encodes a 3-graph K with respect to some spanning semicycle cover $\mathcal{S}$. Then L is spherical.*

Proof. Let C be a circuit in L. By Lemma 6.7, $\beta(C)$ is the blue edge set of a cycle in K. Since $\mathcal{S}$ is spanning, $\beta(C)$ is the blue edge set of a cycle $\sum(\mathcal{U} \cup \mathcal{Y})$ for a set $\mathcal{U}$ of semicycles in $\mathcal{S}$ and a set $\mathcal{Y}$ of red-blue bigons in K. Let $\mathcal{R}$ denote the set of blue-yellow bigons in L that correspond to the semicycles in $\mathcal{U}$. Evidently, $\beta(C) = \beta(\sum(\mathcal{U} \cup \mathcal{Y})) = \beta(\sum(\mathcal{R} \cup \mathcal{Y}))$, and $\sum(\mathcal{R} \cup \mathcal{Y})$ is a cycle in L. By Lemma 6.1, $C = \sum(\mathcal{B} \cup \mathcal{R} \cup \mathcal{Y})$ for some set $\mathcal{B}$ of red-yellow bigons in L. Hence C is a sum of bigons, and we conclude that L is spherical. ■

Let A and B be distinct red-yellow bigons of a 3-graph K. Let c_1 be a yellow edge in A and c_2 a yellow edge in B. Let c_1 join v_1 and w_1 and let c_2 join v_2 and w_2. Let L be the 3-graph obtained from K by deleting c_1 and c_2 and inserting two new yellow edges that join v_1 to v_2 and w_1 to w_2, respectively. We say that L is a 3-graph obtained by *coalescing* A and B. If A and B belong to distinct components of K, then clearly $c(L) = c(K) - 1$.

Lemma 6.10 *Suppose that L is partially congruent to a 3-graph K with respect to a partition $\mathcal{Q}$. Furthermore, suppose that J is obtained from L by coalescing two red-yellow bigons that belong to the same cell in $\mathcal{Q}$. Then J is partially congruent to K. Moreover, if L is faithful to K with respect to a semicycle cover $\mathcal{S}$, then J is faithful to K with respect to $\mathcal{S}$.*

Proof. The first conclusion is immediate. Suppose therefore that L is faithful to K with respect to a semicycle cover $\mathcal{S}$. Let B_1, B_2 be the red-yellow bigons whose coalition yields J. Then, for any $C \in \mathcal{S}$, $|C \cap \partial V B_1|$ and $|C \cap \partial V B_2|$ are both even, and therefore $|C \cap \partial V B|$ is even for all $B \in \mathcal{B}(J)$. Hence J is faithful to K with respect to $\mathcal{S}$. ■

Lemma 6.11 *Let A and B be red-yellow bigons that belong to distinct components of a 3-graph K. Let L denote a 3-graph obtained from K by coalescing A and B. Then $\chi(L) = \chi(K) - 2$. Hence L is spherical if and only if K is spherical.*

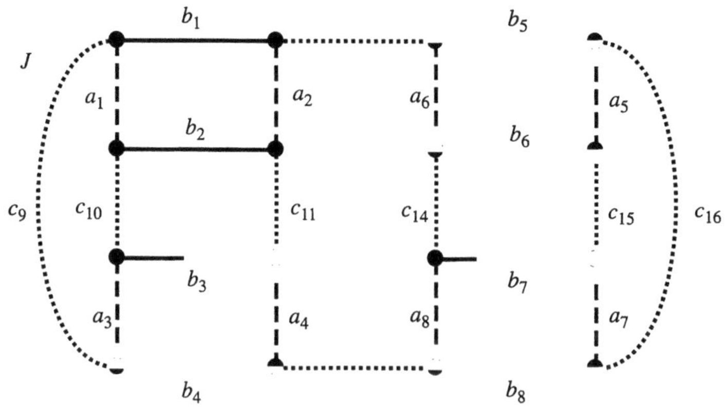

FIGURE 6.6.

Proof. Clearly, $|VL| = |VK|$. However, the number of red-yellow bigons has dropped by 1, as has the number of blue-yellow bigons. Hence

$$\begin{aligned} \chi(L) &= |\mathcal{B}(K)| + |\mathcal{R}(K)| + |\mathcal{Y}(K)| - 2 - \frac{|VK|}{2} \\ &= \chi(K) - 2. \blacksquare \end{aligned}$$

Example 6.7 *Consider the* 3*-graphs K and L of Example 6.5. The* 3*-graph J in Figure 6.6 is a* 3*-graph obtained from L by coalescing* B_2 *and* B_3*. We note that J is a spherical* 3*-graph congruent to K.*

Theorem 6.12 *If there exists a spanning semicycle cover in a* 3*-graph K, then K is congruent to a spherical* 3*-graph.*

Proof. Let L_1 be the 3-graph that encodes K with respect to a spanning semicycle cover $\mathcal{S}$. Let $\mathcal{Q}_1$ be the encoding partition. By Lemma 6.9, L_1 is spherical. Recall that L_1 is partially congruent to K with respect to $\mathcal{Q}_1$ and is faithful to K. If each cell of $\mathcal{Q}_1$ is of cardinality 1, then L_1 is congruent to K and we are done.

Now suppose that there are two distinct red-yellow bigons B_1 and B_2 that belong to a common cell of $\mathcal{Q}_1$. By Lemma 6.8, B_1 and B_2 belong to distinct components of L_1. Let L_2 denote a 3-graph obtained from L_1 by coalescing B_1 and B_2. By Lemma 6.11, L_2 is spherical. By Lemma 6.10, we conclude that L_2 is faithful to K.

Proceeding inductively, we obtain a spherical 3-graph L_n partially congruent to K with respect to a partition $\mathcal{Q}_n$ such that each cell in $\mathcal{Q}_n$ is of cardinality 1. Hence L_n is congruent to K. ■

By combining Theorems 6.6 and 6.12, we obtain the following result.

Theorem 6.13 *A* 3*-graph is congruent to a spherical* 3*-graph if and only if it has a spanning semicycle cover.*

6.5 Mac Lane's theorem

In this section, we specialise Theorem 6.13 to the case of gems to obtain Mac Lane's theorem. The *foundation*, $\mathrm{Fnd}(G)$, of a graph G is the subgraph spanned by the union of the circuits of G. By Theorem 1.14 (p. 9), the complement in EG of its edge set is the set of isthmuses of G.

If C is a cycle in a 3-graph K, then we denote by $N_Y(C)$ the set of all red-blue bigons that meet $\beta(C)$.

Theorem 6.14 (Mac Lane) *A graph is planar if and only if its foundation has a spanning circuit double cover.*

Proof. First, let G be a 2-connected planar graph. Then, by Theorem 5.4 (p. 69) and Corollary 1.37 (p. 21), the face boundaries of a planar imbedding of G constitute a spanning circuit double cover of G. Thus, each block of any planar graph has a spanning circuit double cover. The theorem follows immediately for planar graphs.

Now suppose that G is a graph for which $\mathrm{Fnd}(G)$ has a spanning circuit double cover $\mathcal{C} = (C_1, C_2, \ldots, C_m)$. Let K be a gem corresponding to G. For each $i \in J_m$, let S_i be a semicycle in K that represents C_i, and let $\mathcal{S} = \{S_1, S_2, \ldots, S_m\} \cup \mathcal{S}'$, where $\mathcal{S}'$ is the set of red-blue bigons in K that correspond to the isthmuses in G. Since $S_i + Y$ is a semicycle that also represents C_i, where $i \in J_m$ and $Y \in N_Y(S_i)$, and each red-blue bigon not in $\mathcal{S}'$ belongs to $N_Y(S_i)$ for exactly two values of $i \in J_m$, we may choose each S_i so that the set of intersections of members of $\mathcal{S}$ with $\beta(K)$ is a partition of $\beta(K)$.

We now show that $\mathcal{B}(K) \cup \mathcal{Y}(K) \cup \mathcal{S}$ spans $\mathcal{C}(K)$. Let D be a cycle in K. Suppose that there is a red-blue bigon Z such that both blue edges of Z belong to D. Then $D + Z$ is a sum of circuits in $\mathcal{B}(K) \cup \mathcal{Y}(K) \cup \mathcal{S}$ if and only if D is a sum of circuits in $\mathcal{B}(K) \cup \mathcal{Y}(K) \cup \mathcal{S}$. We therefore assume that D does not contain both blue edges of a red-blue bigon. Since D is a cycle, the number of blue edges of D incident on vertices of a given red-yellow bigon is always even. Therefore, the set of edges of G that correspond to the bigons of $N_Y(D)$ is a cycle C. Since $\mathcal{C}$ spans $\mathcal{C}(G)$, it follows that $C = \sum \mathcal{V}$ for some set $\mathcal{V}$ of components of $\mathcal{C}$. Let $\mathcal{U}$ be the set of semicycles in $\mathcal{S}$ that represent the circuits of $\mathcal{V}$. Let $D_1 = \sum \mathcal{U}$. Let $D_2 = D_1 + \sum \mathcal{U}_3$, where $\mathcal{U}_3$ is the set of red-blue bigons that have both blue edges in D_1. Evidently, D_2 is a cycle such that $N_Y(D_2) = N_Y(D)$ and is a sum of circuits in $\mathcal{Y}(K) \cup \mathcal{S}$. By adding to D_2 bigons in $N_Y(D_2)$, we can obtain a cycle D_3 such that $\beta(D_3) = \beta(D)$. By Lemma 6.1, the fact that D_3 is a sum of circuits in $\mathcal{Y}(K) \cup \mathcal{S}$ implies that D is a sum of circuits in $\mathcal{B}(K) \cup \mathcal{Y}(K) \cup \mathcal{S}$. We conclude that $\mathcal{S}$ is a spanning semicycle cover in K. By Theorem 6.13, K is congruent to a spherical gem K'. Evidently, G underlies K' and hence G is planar. ■

A 3-graph is a 3-*block* if any two blue edges belong to a semicycle. Thus, any 3-graph congruent to a 3-block is also a 3-block. By Theorem 1.26, a gem with at least two red-blue bigons is a 3-block if and only if the underlying graph is

2-connected. The following theorem is closely related to Theorem 6.13 and can be used to strengthen Mac Lane's theorem for 2-connected graphs.

Theorem 6.15 *Let $\mathcal{S}$ be a spanning semicycle cover of a 3-block K. Then K is congruent to a spherical 3-graph L whose blue-yellow bigons are semicycles corresponding to the members of $\mathcal{S}$.*

Proof. Let L_1 be the 3-graph that encodes K with respect to $\mathcal{S}$, and let $\mathcal{Q}_1$ be the encoding partition. Then L_1 is partially congruent to K with respect to $\mathcal{Q}_1$ and faithful to K with respect to $\mathcal{S}$. From Lemma 6.9, L_1 is spherical. Suppose that some cell of $\mathcal{Q}_1$ contains two red-yellow bigons; by Lemma 6.8, they belong to distinct components of L_1. The graph L_2 resulting from their coalition is not a 3-block, as no semicycle in L_2 contains blue edges in distinct components of L_1. By Lemma 6.10, L_2 is partially congruent to K with respect to a partition $\mathcal{Q}_2$ and is faithful to K with respect to $\mathcal{S}$; it is spherical by Lemma 6.11. Proceeding inductively, we reach the contradiction that K is not a 3-block. Therefore, each cell of the encoding partition is of cardinality 1. Hence L_1 is congruent to K. ■

The following corollary is immediate.

Corollary 6.16 *If G is a 2-connected graph with spanning circuit double cover $\mathcal{C}$, then there is a planar imbedding of G for which each component of $\mathcal{C}$ constitutes a face boundary.*

6.6 Whitney's characterisation

Two graphs with a common edge set are *orthogonal* if their cycle spaces are orthogonal complements of each other. A characterisation of planar graphs, due to Whitney (1932), can be expressed in terms of this idea. We present it here, as its proof is a simple consequence of Mac Lane's theorem.

Theorem 6.17 (Whitney) *A graph is planar if and only if another graph is orthogonal to it.*

Proof. Let G be a graph and H a graph orthogonal to it. Then the cocycles of H are the cycles of G. By Theorem 1.32, the vertex cocycles of H span $\mathcal{C}^{\perp}(H) = \mathcal{C}(G)$, and their union, which is the set of links of H, is $\text{Fnd}(G)$. Since each link of H belongs to exactly two vertex cocycles, the set of vertex cocycles of H defines a spanning cycle double cover of $\text{Fnd}(G)$. By applying Corollary 1.37 to each block of $\text{Fnd}(G)$, we deduce that $\text{Fnd}(G)$ has a spanning circuit double cover. Hence G is planar by Mac Lane's theorem.

Conversely, suppose that $(C_1, \ldots, C_n)$ is a spanning cycle double cover $\mathcal{D}$ of the foundation of a graph G. Let H be a graph with vertex set J_n and edge set EG, such that any isthmus of G is a loop in H and any edge e in $\text{Fnd}(G)$ joins distinct vertices i and j in H if and only if $e \in C_i \cap C_j$. Note that

$$\begin{aligned} \partial i &= \bigcup_{j \in J_n - \{i\}} [i, j] \\ &= \bigcup_{j \in J_n - \{i\}} (C_i \cap C_j) \\ &= C_i. \end{aligned}$$

We must show that $\mathcal{C}^{\perp}(H) = \mathcal{C}(G)$. Choose $S \subseteq VH$. Then $\partial S = \sum_{i \in S} \partial i = \sum_{i \in S} C_i$, which is a cycle of G. Conversely, any cycle C of G is a sum of components of $\mathcal{D}$. Let T be the set of subscripts of these components. Then $C = \sum_{i \in T} \partial i = \partial T$, which is a cocycle of H. The proof is complete. ■

7

Kuratowski's Theorem

We move on to prove the famous planarity criterion due to Kuratowski, which characterises planar graphs in terms of forbidden subgraphs.

A simple graph is said to be *complete* if any two distinct vertices are adjacent. For any non-negative integer n, we define K_n to be a complete graph with just n vertices. Any member of the equivalence class $[K_n]$ is then also denoted by K_n.

A simple bipartite graph G is called a *complete* bipartite graph if there exists a non-empty proper subset S of VG such that each vertex in S is adjacent to every vertex in $VG - S$. If $|S| = m$ and $|VG - S| = n$, then we define $K_{m,n} = G$. Thus, $K_{m,n} = K_{n,m}$. Any member of the equivalence class $[K_{m,n}]$ is also denoted by $K_{m,n}$. Despite the terminology, the complete bipartite graph $K_{m,n}$ is not a complete graph if $mn > 1$ – the adjective "complete" here modifies "bipartite graph" rather than "graph".

K_5 and $K_{3,3}$ are drawn in Figure 7.1. The goal of this chapter is to prove that a graph G is non-planar if and only if some subgraph of G is a subdivision of K_5 or $K_{3,3}$.

7.1 Corollaries of Mac Lane's theorem

We begin with several lemmas that are elementary consequences of Mac Lane's theorem.

Lemma 7.1 *Every subgraph of a planar graph is planar.*

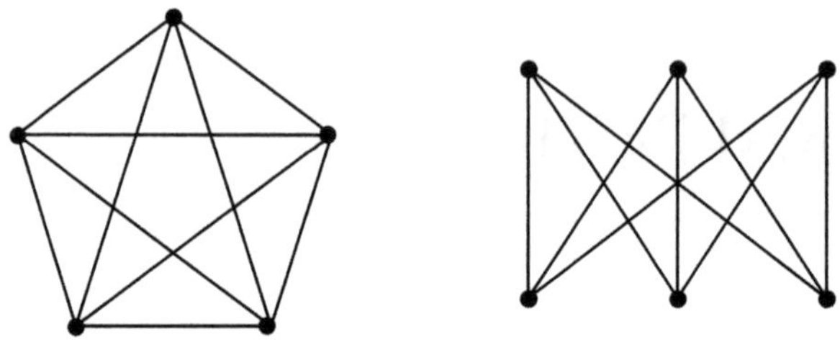

K_5 $K_{3,3}$

FIGURE 7.1.

Proof. Any subgraph of a graph may be formed by the deletion of edges followed by the deletion of isolated vertices. It follows that we need only prove $G-\{e\}$ planar, where e is any edge of a planar graph G. Accordingly, we define $H = G - \{e\}$.

Since G is planar, its foundation has a spanning circuit double cover $\mathcal{D}$. If e is an isthmus, then G and H have a common foundation. Hence H is planar in this case.

Suppose therefore that e is not an isthmus of G. Then e belongs to two components, A and B, of $\mathcal{D}$. Let $C = A + B$, and choose an edge a in Fnd(H). Then a is an edge in Fnd(G). If $a \in A \cap B$, then no circuit of G can contain a but not e, since $\mathcal{D}$ is spanning and the cardinality of the intersection of each component of $\mathcal{D}$ with $\{e, a\}$ is even. In this case, we have the contradiction that a is not in Fnd(H). If $a \in EG - (A \cup B)$, then a belongs to two circuits of $\mathcal{D}$ that are also circuits in H. In the remaining case, we have $a \in C$. Since C is a cycle, C is the union of a set $\mathcal{S}$ of disjoint circuits. The members of $\mathcal{S}$ are circuits of H, since $e \notin C$. Moreover, a belongs to a unique member of $\mathcal{S}$, and also to a unique member of $\mathcal{D}$ distinct from A and B. We conclude that a circuit double cover $\mathcal{E}$ of Fnd(H) is obtained from $\mathcal{D}$ by replacing A and B with the circuits in $\mathcal{S}$.

Finally, choose a circuit D in H. Then D is a circuit in G and hence a sum of components of $\mathcal{D}$. Moreover, $e \notin D$, so that both A and B feature in the sum or neither does. Since $A + B = C = \sum \mathcal{S}$, it follows that D is a sum of components of $\mathcal{E}$. Hence $\mathcal{E}$ spans $\mathcal{C}(H)$. We conclude that H is planar. ■

Combining this result with Theorem 1.27 (p. 17), we obtain the following lemma.

Lemma 7.2 *A graph is planar if and only if each of its blocks is.*

Lemma 7.3 *A graph is planar if and only if its skeleton is.*

Proof. That the skeleton of a planar graph is planar follows immediately from Lemma 7.1.

Conversely, suppose that a graph G is obtained from a planar graph H by adjoining a loop. Let $\mathcal{E}$ be a circuit double cover of Fnd(G) obtained from a spanning circuit double cover of Fnd(H) by adjoining the loop twice. Then $\mathcal{E}$ is spanning, since any circuit of G other than the loop is a circuit of H.

Finally, suppose that G is obtained from the planar graph H by adjoining an edge e whose ends are adjacent in H. Let a be an edge of H with the same ends as e. Let $\mathcal{D}$ be a spanning circuit double cover of $\mathrm{Fnd}(H)$. We distinguish two cases.

Case 1. Suppose that a is an isthmus of H. Let $\mathcal{E}$ be the circuit double cover of $\mathrm{Fnd}(G)$ obtained from $\mathcal{D}$ by adjoining the circuit $\{e, a\}$ twice. Then $\mathcal{E}$ is spanning, since any circuit of G other than $\{e, a\}$ is a circuit of H.

Case 2. Suppose that a is not an isthmus of H. Let A and B be the components of $\mathcal{D}$ that contain a. Let $\mathcal{E}$ be the circuit double cover of $\mathrm{Fnd}(G)$ obtained from $\mathcal{D}$ by replacing B with $B + \{e, a\}$ and adjoining $\{e, a\}$. Any circuit of G, other than $\{e, a\}$, that is not a circuit of H is of the form $C + \{e, a\}$ for some circuit C of H. Hence $\mathcal{E}$ is spanning.

In all cases, we conclude from Mac Lane's theorem that G is planar. ∎

Lemma 7.4 *Let G be a planar graph and let e be a link of G. Then G_e is planar.*

Proof. Let $\mathcal{D}$ be a spanning circuit double cover of $\mathrm{Fnd}(G)$. The non-empty contractions in G_e of the components of $\mathcal{D}$ are circuits of G_e, and these circuits constitute a circuit double cover $\mathcal{E}$ of $\mathrm{Fnd}(G_e)$. Finally, choose any circuit C in G_e. There is a circuit D in G that contracts to C in G_e. Thus, D is a sum of components of $\mathcal{D}$, and C is the sum of the contractions of these components in G_e. Hence $\mathcal{E}$ is spanning. ∎

Lemma 7.5 *Let H be a subdivision of a graph G. Then H is planar if and only if G is.*

Proof. We may assume that H is an elementary subdivision of G. If H is planar, then so is G by Lemma 7.4. Conversely, let $\mathcal{E}$ be a spanning circuit double cover of $\mathrm{Fnd}(G)$. Each component of $\mathcal{E}$ is the contraction in G of a circuit of H, and these circuits constitute a spanning circuit double cover of $\mathrm{Fnd}(H)$. ∎

Lemma 7.6 *Any non-planar graph has at least five vertices.*

Proof. By Lemma 7.3, we need consider only simple graphs. By Lemma 7.1 and Mac Lane's theorem, it therefore suffices to give a spanning circuit double cover of K_4. But the construction is immediate: The four triangles of K_4 constitute a circuit double cover, and the sum of any two of them is a square. ∎

We now strengthen Lemma 7.2.

Theorem 7.7 *Any non-planar graph includes a subdivision of a 3-connected non-planar graph.*

Proof. Let G be a non-planar graph that does not include a subdivision of a 3-connected non-planar graph. We may choose G to have as few vertices as possible. Then G is a block, by Lemma 7.2.

Since G is not 3-connected, there are distinct vertices u and v such that $G - \{u, v\}$ is not connected. Let K and L be components of $G - \{u, v\}$. Since G is 2-connected, K and L both contain a vertex adjacent to u and a vertex adjacent to v. Thus, the graphs $G[VK \cup \{u, v\}]$ and $G[VL \cup \{u, v\}]$ are connected. Let K' be the graph obtained from $G[VK \cup \{u, v\}]$ by the adjunction of an edge e joining u and v. Thus $e \in E(\text{Fnd}(K'))$. Similarly, let L' be the graph obtained from $G[VL \cup \{u, v\}]$ by the adjunction of an edge a joining u and v. Note that $|VK'| < |VG|$ and $|VL'| < |VG|$. Thus, if K' were non-planar then K' would include a subdivision of a 3-connected non-planar graph. Letting P be a path in $L' - \{a\}$ joining u and v, we should obtain the contradiction that the subgraph $G[(EK' - \{e\}) \cup P]$ of G includes a subdivision of a 3-connected non-planar graph. We infer that K' must be planar. Similarly L' is planar. Moreover, Lemma 7.3 implies that K'' and L'' are also planar, where K'' and L'' are obtained from K' and L', respectively, by the adjunction of edges e' and a', respectively, joining u and v.

Now let G' be the graph obtained from G by the adjunction of two edges, c and d, joining u and v. We shall prove that G' is planar.

First, we construct a spanning circuit double cover $\mathcal{E}$ for K'. We then obtain a circuit double cover $\mathcal{E}'$ of K'' by first replacing e by e' in one of the components of $\mathcal{E}$, and then adjoining the digon $\{e, e'\}$. $\mathcal{E}'$ is clearly spanning, for any circuit of K'' other than $\{e, e'\}$ that is not a circuit of K' is of the form $C + \{e, e'\}$ for some circuit C of K'. Similarly, we construct a spanning circuit double cover $\mathcal{F}'$ of L'' that contains the digon $\{a, a'\}$ as a component.

We now use $\mathcal{E}'$ and $\mathcal{F}'$ to construct a circuit double cover $\mathcal{D}$ of G'. First, we remove the components $\{e, e'\}$ and $\{a, a'\}$ from $\mathcal{E}'$ and $\mathcal{F}'$, respectively. We then obtain $\mathcal{D}$ by replacing e and e' by c and d, respectively, in the remaining components of $\mathcal{E}'$ and replacing a and a' by c and d, respectively, in the remaining components of $\mathcal{F}'$.

It remains to show that $\mathcal{D}$ is spanning. Note first that the graph $G'[VK']$ is isomorphic to K''. Any circuit of K'' can be written as a sum of components of the spanning double cover $\mathcal{E}'$ distinct from $\{e, e'\}$. Hence, any circuit of $G'[VK']$ is a sum of components of $\mathcal{D}$. A similar statement holds for circuits of $G'[VL']$. Finally, let D be a circuit such that VD meets both VK and VL. Then D includes a unique path Q that joins u and v and is included in EK', and also includes a unique path R that joins u and v and is included in EL'. Moreover, $Q \cup \{c\}$ and $R \cup \{d\}$ are circuits. We call these circuits A and B, respectively. Then $D = A + B + \{c, d\}$, and A, B, and $\{c, d\}$ are circuits of $G'[VK']$ or $G'[VL']$. Hence $\mathcal{D}$ is spanning.

We conclude that G' is planar. By Lemma 7.1, we now obtain the contradiction that G is planar. ■

We also require the following technical result. Intuitively, it amounts to the following assertion. Let I and J be two planar imbeddings of planar graphs. Then the identification of a face boundary in I with a face boundary in J yields a planar graph.

Lemma 7.8 *Let $\mathcal{S}$ and $\mathcal{T}$ be sets of circuits in a graph G, and suppose that $\bigcup(\mathcal{S} \cup \mathcal{T}) = EG$, that $\bigcup \mathcal{S} \cap \bigcup \mathcal{T}$ is a circuit $C \notin \mathcal{S} \cup \mathcal{T}$, and that $VG[\bigcup \mathcal{S}] \cap VG[\bigcup \mathcal{T}] = VC$. Suppose also that $\mathcal{S} \cup \{C\}$ and $\mathcal{T} \cup \{C\}$ are spanning circuit double covers of the graphs $G[\bigcup \mathcal{S}]$ and $G[\bigcup \mathcal{T}]$, respectively. Then G is planar.*

Proof. Note that $\mathcal{S} \cap \mathcal{T} = \emptyset$, since $\bigcup\mathcal{S} \cap \bigcup\mathcal{T} = C \notin \mathcal{S} \cup \mathcal{T}$.

Choose $e \in EG$. If $e \in C$, then $e \in \bigcup\mathcal{S}$, since $\bigcup\mathcal{S} \cap \bigcup\mathcal{T} = C$. Since $C \notin \mathcal{S}$ and $\mathcal{S} \cup \{C\}$ is a circuit double cover of $G[\bigcup\mathcal{S}]$, it follows that e is contained in a unique member of $\mathcal{S}$. Similarly, e is contained in a unique member of $\mathcal{T}$, and hence in precisely two members of $\mathcal{S} \cup \mathcal{T}$ since $\mathcal{S} \cap \mathcal{T} = \emptyset$. If $e \in \bigcup\mathcal{S} - C$, then $e \notin \bigcup\mathcal{T}$. In this case, e belongs to just two members of $\mathcal{S}$ but to no member of $\mathcal{T}$. Similarly, if $e \in \bigcup\mathcal{T} - C$, then e is contained in just two circuits of $\mathcal{S} \cup \mathcal{T}$. Hence $\mathcal{S} \cup \mathcal{T}$ is a circuit double cover, since $\bigcup(\mathcal{S} \cup \mathcal{T}) = EG$.

Choose a circuit D of G.

Case 1. Suppose that $D \subseteq \bigcup\mathcal{S}$. Since $\mathcal{S} \cup \{C\}$ spans $\mathcal{C}(G[\bigcup\mathcal{S}])$, D is a sum of elements of $\mathcal{S} \cup \{C\}$. Since $\mathcal{S} \cup \{C\}$ is a double cover, we have $C + \sum\mathcal{S} = \emptyset$, so that $C = \sum\mathcal{S}$. Hence D is a sum of circuits in $\mathcal{S}$.

Case 2. Similarly, D is a sum of circuits in $\mathcal{T}$ if $D \subseteq \bigcup\mathcal{T}$.

Case 3. In the remaining case, we use induction on $|D-C|$. If $|D-C| = 0$, then $D = C = \sum\mathcal{S}$. Suppose therefore that $D-C \neq \emptyset$ and, as an inductive hypothesis, that any circuit D' satisfying $|D' - C| < |D - C|$ is a sum of circuits in $\mathcal{S} \cup \mathcal{T}$. Then $D-C$ includes a path P, of minimum length, joining two vertices of C. The hypothesis that $VG[\bigcup\mathcal{S}] \cap VG[\bigcup\mathcal{T}] = VC$ implies that $P \subset \bigcup\mathcal{S}$ or $P \subset \bigcup\mathcal{T}$ since no internal vertex of the path P can belong to VC. We may suppose without loss of generality that $P \subset \bigcup\mathcal{S}$. Let x, y be the ends of P. Thus $x \in VC$ and $y \in VC$. C includes a path Q joining x and y, and $P \cup Q$ is a circuit. By Case 1, this circuit is a sum of circuits in $\mathcal{S}$. Moreover, the cycle $X = D + (P \cup Q)$ is included in $C \cup D$, and the inclusion is proper since $P \cap X = \emptyset$. Thus X is a union of disjoint circuits, each a sum of elements of $\mathcal{S} \cup \mathcal{T}$ by the inductive hypothesis. Since D is the sum of these circuits and $P \cup Q$, we conclude that $\mathcal{S} \cup \mathcal{T}$ spans $\mathcal{C}(G)$.

It follows that $\mathcal{S} \cup \mathcal{T}$ is a spanning circuit double cover for G and hence, by Mac Lane's theorem, that G is planar. ∎

7.2 Kuratowski's theorem

Mac Lane's theorem can be used to prove K_5 and $K_{3,3}$ non-planar. Indeed, if K_5 were planar, then by Corollary 1.36 (p. 21) the number of components of any spanning circuit double cover would be $|EK_5| - |VK_5| + 2 = 10 - 5 + 2 = 7$. The length of each component of the double cover would be at least 3 (since K_5 is simple). The sum of the lengths of these circuits would therefore be at least $7 \cdot 3 = 21$, which exceeds $20 = 2|EK_5|$. This contradiction establishes the non-planarity of K_5. Likewise, if $K_{3,3}$ were planar, there would be $9 - 6 + 2 = 5$ components of a spanning circuit double cover. The length of each would be 4 or more, since $K_{3,3}$ is simple and bipartite. The sum of their lengths would therefore be at least $5 \cdot 4 = 20 > 18 = 2|EK_{3,3}|$. Hence $K_{3,3}$ is non-planar.

Using Lemma 7.1 and Lemma 7.5, we conclude that if a graph G has a subgraph that is a subdivision of K_5 or $K_{3,3}$, then G is non-planar. Kuratowski's theorem is the converse of this assertion. We prepare for its proof with the following lemma.

A path P in a graph G joining two vertices of degree 3 or more is called an *ear* if all its internal vertices have degree 2 in G.

Lemma 7.9 *Let G be a graph that includes no subdivision of K_5 or $K_{3,3}$. Then G_e includes no such subdivision for each link $e \in EG$.*

Proof. Let H be a subgraph of G_e. We shall prove by contradiction that H is not a subdivision of K_5 or $K_{3,3}$.

Suppose the contrary. Let e join vertices u and v in G. Thus $\{u, v\} \in VH$. Let S be a set of edges of G that contracts to EH in G_e, and let L be the skeleton of the graph $G[S \cup \{e\}]$. If $\deg_L(u) = 1$, then $\partial_L u = \{e\}$. In this case, we find that $G[S]$ is isomorphic to H, in contradiction to the hypothesis that G includes no subdivision of K_5 or $K_{3,3}$. If $\deg_L(u) = 2$, then similarly we have the contradiction that $G[S \cup \{e\}]$ is isomorphic to a subdivision of H. Hence $\deg_L(u) \geq 3$, and by the same argument $\deg_L(v) \geq 3$. But

$$\deg_H(\{u, v\}) = \deg_L(u) + \deg_L(v) - 2,$$

and since no vertex of H is of degree greater than 4, we deduce that $\deg_L(u) = \deg_L(v) = 3$, that $\deg_H(\{u, v\}) = 4$, and hence that H is a subdivision of K_5. Therefore L has just four distinct vertices w, x, y, z of degree 4. Without loss of generality, we may assume that L includes ears joining u to w and to x, and ears joining v to y and to z. L also includes ears joining any two of w, x, y, z. (See Figure 7.2.) We now have the contradiction that L includes a subdivision J of $K_{3,3}$: We have $EJ = EL - (P \cup Q)$, where P and Q are ears in L joining w to x and y to z, respectively. ■

We also need the following lemma, which is due to Thomassen (1981).

Lemma 7.10 *Any simple 3-connected graph G with $|VG| \geq 5$ has a link e such that the skeleton of G_e is 3-connected.*

Proof. Suppose that the lemma is false. Then there exists a simple 3-connected graph G, with $|VG| \geq 5$, such that for each link $e \in EG$ the skeleton of the

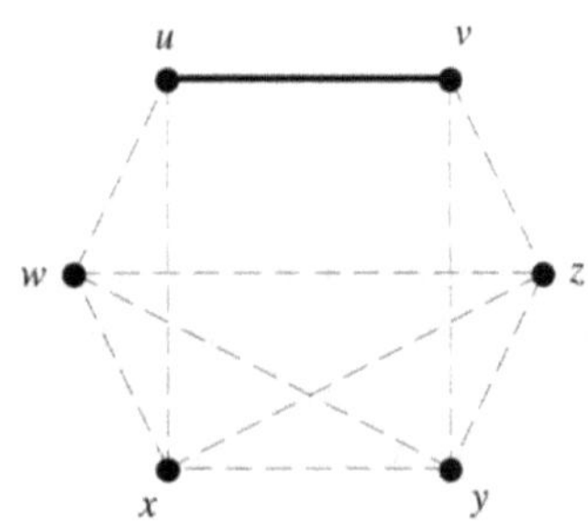

FIGURE 7.2.

graph G_e is not 3-connected. The 3-connectedness of G implies that the skeleton of G_e must be 2-connected. Moreover, since $|VG_e| \geq 4$, G must contain a vertex $w \notin VG[\{e\}]$ such that $G_e - \{\{w\}, \{u, v\}\}$ is not connected, where u and v are the ends of e. Thus $G - \{u, v, w\}$ is not connected. We may assume (e, w) to be chosen to maximise $|VH|$, where H is a component of $G - \{u, v, w\}$ with the largest number of vertices.

Let X be another component of $G - \{u, v, w\}$. The 3-connectedness of G implies that w is adjacent to a vertex $x \in VX$. Since w and x are joined by a link, there exists a vertex $y \in VG - \{w, x\}$ such that $G - \{w, x, y\}$ is not connected.

We shall show that $G - \{w, y\}$ is not connected, thereby contradicting the 3-connectedness of G. Let $K = G[VH \cup \{u, v\}]$. Since G is 3-connected, u and v must be adjacent to vertices of H, so that K is connected. If $y \notin VK$, then K is a subgraph of some component of $G - \{w, x, y\}$, in contradiction to the choice of (e, w). Thus $y \in VK$. The same argument shows that $K - \{y\}$ is not connected, since $|VK - \{y\}| = |VK| - 1 > |VH|$. Hence $y \notin \{u, v\}$; moreover, u and v are in the same component of $K - \{y\}$. Choose $z \in VZ$, where Z is another component of $K - \{y\}$. Since x and z are in distinct components of $G - \{u, v, w\}$, every path in G joining z and x passes through u, v, or w. But every path in G joining z to u or v passes through w or y. Hence x and z are in distinct components of $G - \{w, y\}$. ■

Theorem 7.11 (Kuratowski) *A graph is non-planar if and only if it includes no subdivision of K_5 or $K_{3,3}$.*

Proof. We have already seen that any graph that includes a subdivision of K_5 or $K_{3,3}$ is non-planar.

Conversely, suppose that there exists a non-planar graph G that includes no subdivision of K_5 or $K_{3,3}$. Choose G to minimise $|VG| + |EG|$. Then $|VG| \geq 5$ and G is simple. By Theorem 7.7 and Corollary 7.5 (p. 99), it follows that G is 3-connected. By Lemma 7.10, it therefore has a link e for which the skeleton H of the contraction G_e is 3-connected. Thus, if e joins vertices u and v, then $\{u, v\}$ is a vertex a of H of degree at least 3. Note also that, by Lemma 7.9, H includes no subdivision of K_5 or $K_{3,3}$. By the choice of G, it follows that H is planar. Thus H has a spanning circuit double cover $\mathcal{E}$.

Let $\mathcal{D}$ be the set of components of $\mathcal{E}$ that pass through a. We shall show that $\sum \mathcal{D}$ is a circuit. Let $C = \sum \mathcal{D}$. Let $\mathcal{S} = \mathcal{E} - \mathcal{D}$. By Corollary 1.36 (p.21) we have $\dim \mathcal{C}(H) = |\mathcal{E}| - 1$, and since $\sum \mathcal{E} = \emptyset$ and $\mathcal{E}$ spans $\mathcal{C}(H)$ it follows that the members of any proper subset of $\mathcal{E}$ are linearly independent. Therefore, the circuits in $\mathcal{S}$ are linearly independent. Since the elements of $\mathcal{E}$ are circuits, we also have

$$
\begin{aligned}
|\mathcal{S}| &= |\mathcal{E}| - |\mathcal{D}| \\
&= \dim \mathcal{C}(H) + 1 - \deg(a) \\
&= |EH| - |VH| - \deg(a) + 2 \\
&= |EH'| - (|VH'| + 1) + 2
\end{aligned}
$$

$$= |EH'| - |VH'| + 1$$
$$= \dim \mathcal{C}(H'),$$

where $H' = H - \{a\} = G - \{u, v\}$, and so $\mathcal{S}$ forms a basis of $\mathcal{C}(H')$. Since H is 3-connected, H' must be 2-connected, and it follows that each of its edges belongs to a circuit in $\mathcal{S}$. If C were not a circuit, then $C \notin \mathcal{S}$ and we deduce that $\mathcal{S} \cup \{C\}$ would be a cycle double cover of H', since $\sum \mathcal{S} = \sum \mathcal{E} + \sum \mathcal{D} = \sum \mathcal{D} = C$. Corollary 1.37 now yields the contradiction that C would be a circuit.

Now that we have established that C is a circuit, our next goal is to show that each edge of ∂a is incident on a vertex of VC. Let f be an edge joining a to a vertex b. This edge belongs to just two circuits, A and B, of $\mathcal{D}$. Any edge joining b to a vertex of $VH - \{a\}$ is an edge of H' and therefore belongs to $\bigcup \mathcal{S}$ since H' is 2-connected and $\mathcal{S}$ spans $\mathcal{C}(H')$. Thus f is the only edge of ∂b belonging to $A \cap B$. Some edge incident on b therefore belongs to $A - B$ and to a circuit of $\mathcal{S}$, and hence to $\sum \mathcal{S} = C$. Thus $b \in VC$.

We now turn our attention back to G. If VC contains distinct vertices x, y, z adjacent to both u and v, then we have the contradiction that the graph

$$G[C \cup \{e\} \cup [\{u, v\}, \{x, y, z\}]]$$

is a subdivision of K_5 in G. (See Figure 7.3a.) Next, suppose that VC contains distinct vertices w and x adjacent to u, and that $VC - \{w, x\}$ contains distinct vertices y and z adjacent to v, where $y \in VC_z[w, x]$. Then we have the contradiction that

$$G[C \cup \{e\} \cup [\{u, v\}, \{w, x, y, z\}]]$$

is a subdivision of $K_{3,3}$. (See Figure 7.3b.)

We conclude that there are disjoint paths P and Q included in C such that VP and VQ contain all the vertices of VC adjacent to u and v, respectively. Let $P_1, P_2, \ldots, P_m$ be the ears in the graph $G[C \cup \partial u]$ that are included in P. Then $P_i \cup [VP_i, \{u\}]$ is a circuit, C_i, for each i. Similarly, if $Q_1, Q_2, \ldots, Q_n$ are the ears in the graph $G[C \cup \partial v]$ that are included in Q, then $Q_i \cup [VQ_i, \{v\}]$ is a circuit, D_i, for each i. Let $\mathcal{C} = \{C_1, C_2, \ldots, C_m\}$ and $\mathcal{D} = \{D_1, D_2, \ldots, D_n\}$.

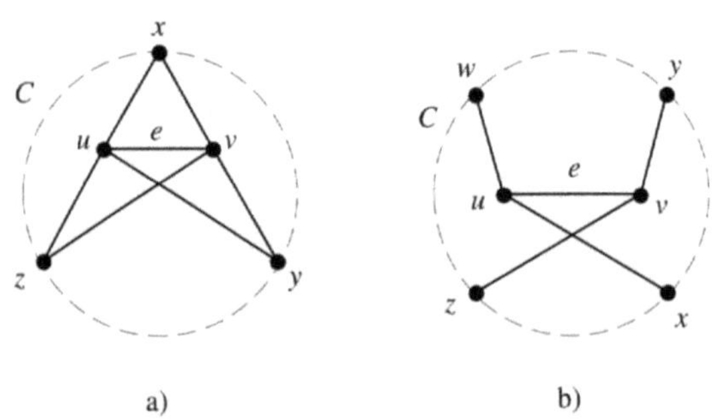

FIGURE 7.3.

Clearly, $C + \sum \mathcal{C} + \sum \mathcal{D}$ is also a circuit, L, and passes through u and v. There are two paths M and N that are included in L and join u and v. Let $M' = M \cup \{e\}$ and $N' = N \cup \{e\}$. Then M' and N' are circuits, and $M' + N' = M + N = L$. Moreover, $M' + N' + \sum \mathcal{C} + \sum \mathcal{D} = C$. Thus, if we define $\mathcal{T} = \mathcal{C} \cup \mathcal{D} \cup \{M', N'\}$, then $\mathcal{T} \cup \{C\}$ is a circuit double cover of $G[\bigcup \mathcal{T}] = G[C \cup \partial u \cup \partial v]$. The circuits of $\mathcal{T}$ are linearly independent since $M' + N' \neq \emptyset$ and any $\mathcal{T}' \subseteq \mathcal{T}$ which meets $\mathcal{C} \cup \mathcal{D}$ has the property that $\sum \mathcal{T}'$ meets C. Note also that

$$\begin{aligned}\dim \mathcal{C}(G[\bigcup \mathcal{T}]) &= |C| + m + n + 3 - (|C| + 2) + 1\\ &= m + n + 2\\ &= |\mathcal{T}|,\end{aligned}$$

and so $\mathcal{T} \cup \{C\}$ is a spanning circuit double cover of $G[\bigcup \mathcal{T}]$. Hence, this graph is planar. Since $G[\bigcup \mathcal{T}]$ is planar while G is not, some vertex of VC must be of degree greater than 2 in the graph $G - \{u, v\} = H'$. Since $\sum \mathcal{S} = C$, it follows that $\mathcal{S} \cup \{C\}$ is a spanning circuit double cover of $H' = G[\bigcup \mathcal{S}]$, and so we deduce from Lemma 1.38 that $C \notin \mathcal{S}$. Moreover $\bigcup \mathcal{S} \cap \bigcup \mathcal{T} = C$. We also have $VG[\bigcup \mathcal{S}] = VG - \{u, v\}$ and $VG[\bigcup \mathcal{T}] = VC \cup \{u, v\}$, so that the intersection of these sets is VC. An appeal to Lemma 7.8 now reveals the planarity of G, in contradiction to hypothesis. ■

An analogous theorem for graphs imbeddable in the projective plane has been proved by Archdeacon (1980, 1981). It transpires that the graphs K_5 and $K_{3,3}$ are replaced by a set of 103 graphs.

It may be remarked at this point that graphs that contain a subdivision of K_5 but no subdivision of $K_{3,3}$ have a very special structure. The details are specified in the following theorem.

Theorem 7.12 *Let G be a graph that contains a subdivision H of K_5 but no subdivision of $K_{3,3}$. Let v, w, x, y, z be the vertices of degree 4 in H. Let a be an internal vertex of the path P in $H - \{x, y, z\}$ joining v and w. Then any path in G joining a to a vertex of $VH - VP$ must have v or w as an internal vertex.*

Proof. Suppose that there is a path R in G joining a to a vertex $b \in VH - VP$ and having no internal vertex in VH. For each pair $\{c, d\} \subset \{v, w, x, y, z\}$ we denote by Q_{cd} the ear in H joining c and d.

Without loss of generality, we confine our attention to three possible choices for b:

(1) $b = x$,

(2) b is an internal vertex of Q_{wx},

(3) b is an internal vertex of Q_{xy}.

(See Figure 7.4.) In each case, respectively, the following graphs are subdivisions of $K_{3,3}$:

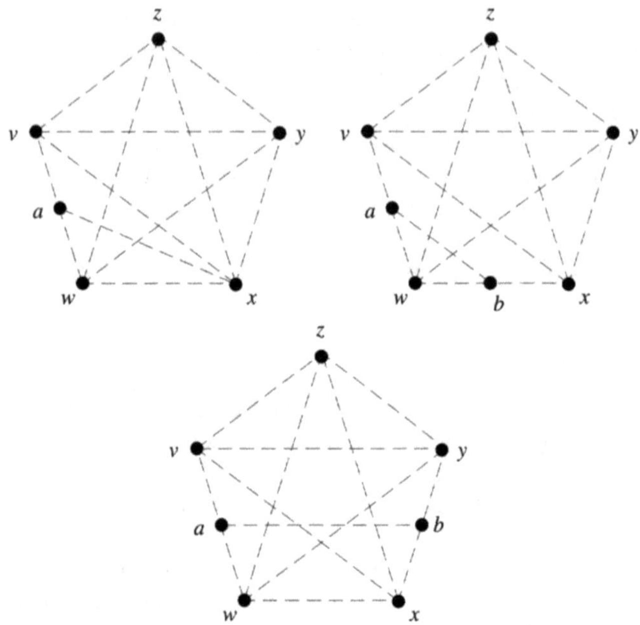

FIGURE 7.4.

(1) $G[(R \cup EH) - (Q_{wx} \cup Q_{vx} \cup Q_{yz})]$,

(2) $G[(R \cup EH) - (Q_{wx}[w, b] \cup Q_{vx} \cup Q_{yz})]$,

(3) $G[(R \cup EH) - (Q_{vz} \cup Q_{wz} \cup Q_{xz} \cup Q_{yz})]$.

Each case therefore contradicts a hypothesis, and so the theorem follows. ■

7.3 Wagner's theorem

The graphs K_5 and $K_{3,3}$ feature in another characterisation of planar graphs, due to Wagner (1937). A *minor* of a graph is a subgraph of a contraction.

Lemma 7.13 *If a graph G contains a subdivision J of a graph H with no loops, then H is a minor of G.*

Proof. It suffices to treat the case where J is an elementary subdivision of H. Then, for some vertex $v \in VJ$ of degree 2 in J, the graph H is obtained from $J - \{v\}$ by adding an edge joining the two vertices x and y adjacent to v. Let G' be the contraction of G determined by the partition of VG, of cardinality $|VG| - 1$, containing the cell $\{v, x\}$. Then H is a subgraph of G' and so a minor of G. ■

Theorem 7.14 (Wagner) *A graph is planar if and only if neither K_5 nor $K_{3,3}$ is a minor of it.*

Proof. If G is a non-planar graph, the theorem follows by Kuratowski's theorem and Lemma 7.13.

Conversely, suppose that G is a graph of which K_5 or $K_{3,3}$ is a minor.

Case 1. Suppose that $K_{3,3}$ is a minor of G. Then there are six mutually disjoint subsets $V_1, V_2, V_3, W_1, W_2, W_3$ of VG such that $G[V_i]$, $G[W_j]$, and $G[V_i \cup W_j]$ are connected whenever $(i, j) \in J_3 \times J_3$. For each $(i, j) \in J_3 \times J_3$ some vertex $v_{ij} \in V_i$ is joined by an edge e_{ij} to a vertex $w_{ji} \in W_j$. Since $G[V_i]$ is connected, V_i contains a path P_{i12} joining v_{i1} and v_{i2}, and a path P_{i13} joining v_{i1} and v_{i3}. (See Figure 7.5.) Since $VP_{i12} \cap VP_{i13} \neq \emptyset$, we may select the vertex $v_i \in VP_{i12} \cap VP_{i13}$ that minimises $|P_{i13}[v_i, v_{i3}]|$. Define $E_i = P_{i12} \cup P_{i13}[v_i, v_{i3}]$. Similarly, for each $j \in J_3$ we may define Q_{j12}, Q_{j13}, w_j, and F_j analogously to the way in which P_{i12}, P_{i13}, v_i, and E_i, respectively, have been defined. Then

$$E_1 \cup E_2 \cup E_3 \cup F_1 \cup F_2 \cup F_3 \cup \{e_{11}, e_{12}, e_{13}, e_{21}, e_{22}, e_{23}, e_{31}, e_{32}, e_{33}\}$$

induces a subdivision of $K_{3,3}$. (See Figure 7.6.) Hence G is non-planar.

Case 2. Suppose that K_5 is a minor of G. Then there are five mutually disjoint subsets V_1, V_2, V_3, V_4, V_5 of VG such that $G[V_i]$ and $G[V_i \cup V_j]$ are connected whenever $(i, j) \in J_5 \times J_5$. For each $i \in J_5$ and $j \in J_5 - \{i\}$, there exists an edge joining a vertex $v_{ij} \in V_i$ to a vertex $v_{ji} \in V_j$. We denote this edge by e_{ij} if $i < j$ and by e_{ji} otherwise. Let $J_5 - \{i\} = \{a, b, c, d\}$, where $a < b < c < d$. Then $G[V_i]$ contains a path P_{iab} joining v_{ia} to v_{ib} and a path P_{iac} joining v_{ia} to v_{ic}. We define w_{ic} to be the vertex of $VP_{iab} \cap VP_{iac}$ that minimises $|P_{iac}[w_{ic}, v_{ic}]|$. $G[V_i]$ also contains a path P_{iad} joining v_{id} to w_{ic}. Define w_{id} to be the vertex of $VP_{iad} \cap (VP_{iab} \cup VP_{iac}[w_{ic}, v_{ic}])$ that minimises $|P_{iad}[w_{id}, v_{id}]|$. Define

$$G_i = G[P_{iab} \cup P_{iac}[w_{ic}, v_{ic}] \cup P_{iad}[w_{id}, v_{id}]],$$

and $E_i = EG_i$. (See Figure 7.7.) Thus G_i has a vertex of degree 4 if $w_{ic} = w_{id}$ but not otherwise.

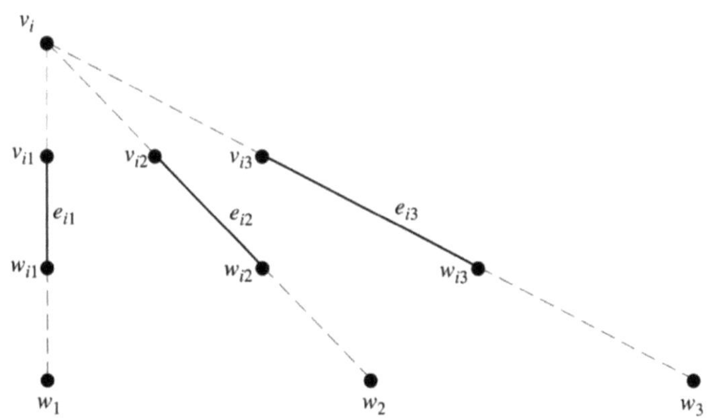

FIGURE 7.5.

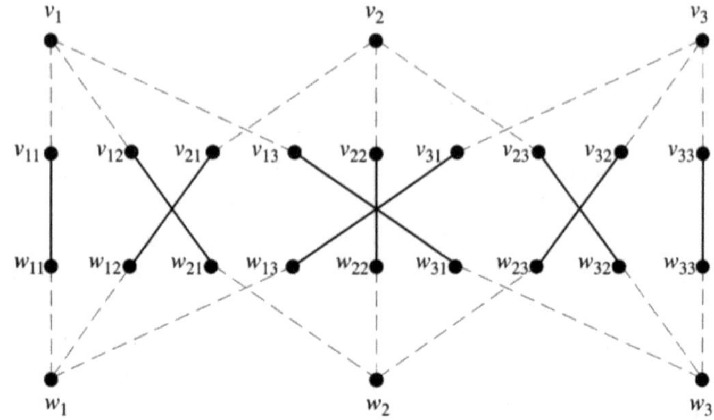

FIGURE 7.6.

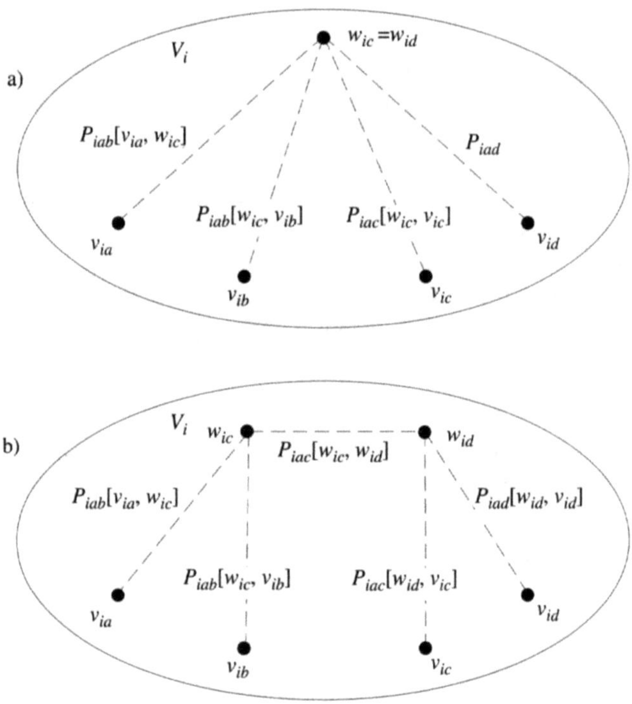

FIGURE 7.7.

If each of G_1, G_2, G_3, G_4, G_5 has a vertex of degree 4, then

$$E_1 \cup E_2 \cup E_3 \cup E_4 \cup E_5 \cup \{e_{12}, e_{13}, e_{14}, e_{15}, e_{23}, e_{24}, e_{25}, e_{34}, e_{35}, e_{45}\}$$

induces a subdivision of K_5, and so G is non-planar. We may therefore suppose without loss of generality that G_1 has no such vertex. Furthermore, without loss of generality, we assume that $w_{15} \in VP_{124}[w_{14}, v_{14}]$. Now $G[V_2]$ contains a path

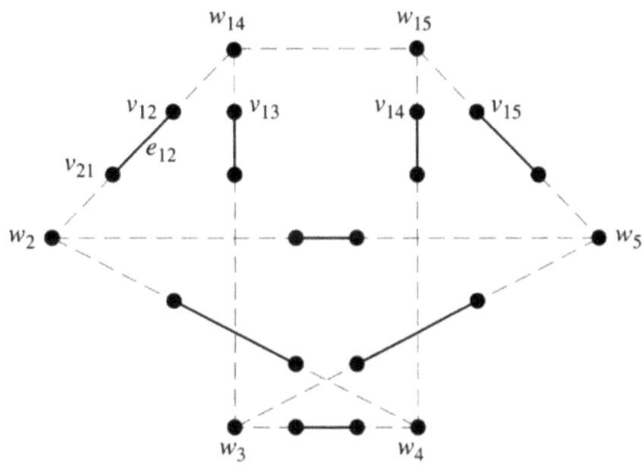

FIGURE 7.8.

Q_{245} joining v_{24} to v_{25} and a path Q_{241} joining v_{24} to v_{21}. We select the vertex $w_2 \in VQ_{245} \cap VQ_{241}$ that minimises $|VQ_{241}[w_2, v_{21}]|$ and define $F_2 = Q_{245} \cup Q_{241}[w_2, v_{21}]$. Define F_3 similarly, replacing the subscript 2 by 3 throughout, and then define F_4 and F_5 analogously, with the ordered pairs (V_2, V_3) and (V_4, V_5) interchanged. Then

$$E_1 \cup F_2 \cup F_3 \cup F_4 \cup F_5 \cup \{e_{12}, e_{13}, e_{14}, e_{15}, e_{24}, e_{25}, e_{34}, e_{35}\}$$

induces a subdivision of $K_{3,3}$, and G is non-planar. (See Figure 7.8.) ■

8
Duality

The concept of a gem, as we have seen, corresponds closely to the idea of an imbedding of a map but is much simpler conceptually. It will also prove to be of theoretical assistance to us in this chapter. We use it to illuminate the notion of duality, which manifests itself topologically as a relationship between vertices and faces. Following the treatment in Lins (1982), we show that this relationship is merely the tip of the iceberg – there are actually three dualities. This work leads naturally to the principal edge tripartition of a graph, studied by Rosenstiehl and Read (1978), and culminates in another characterisation of planar graphs.

8.1 Duals

Let (H, V, F) be a gem K. H is a subgraph of a 4-valent graph L in which each pair of vertices in every bisquare of K is joined. Let $Z = EL - EH$. Thus $H = L - Z$. Z is a fourth *colour*, and its edges will be deemed *green*. We will call (L, V, F, Z) the *jewel* associated with K. Thus there is an obvious 1:1 correspondence between gems and jewels. The ordered 4-tuples (L, F, V, Z), (L, Z, F, V), (L, F, Z, V), (L, Z, V, F), and (L, V, Z, F) are also jewels associated with gems. We may classify the gems with which these jewels are associated as follows. We call $(L - Z, F, V)$ the *dual*, K^D, of K and $(L - V, Z, F)$ the *skew*, K^S, of K. $(L - V, F, Z)$ is then the dual, K^{SD}, of K^S, and $(L - F, Z, V)$ is the skew, K^{DS}, of K^D. Finally, the dual, K^{DSD}, of K^{DS} is $(L - F, V, Z)$, and this gem is the *phial* of K. In order to study the relationship of these gems to K, it therefore suffices to study K^D and K^S. Observe that, given K, one obtains K^D by interchanging the roles of V and

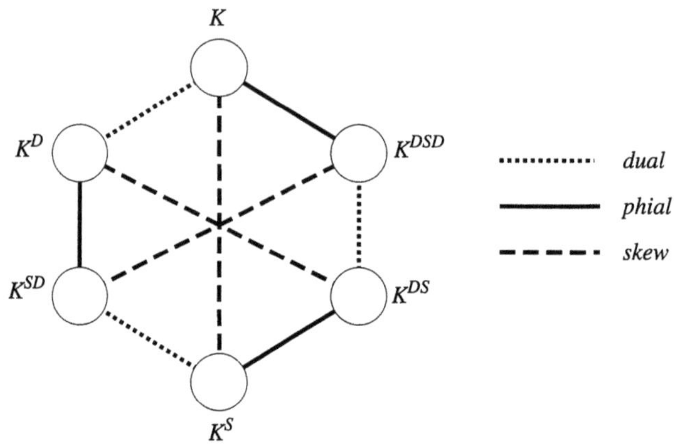

FIGURE 8.1.

F, K^S by interchanging the roles of V and Z, and K^{DSD}, by interchanging the roles of F and Z. These interchanges constitute what Lins calls the gem *dualities*. The structure of the gem dualities is shown in Figure 8.1. Observe that the phial of K is also the skew of the dual of the skew of K. For this reason, we also denote it by K^{SDS}. K^D is obtained from K by interchanging the colours red and blue. Hence $K^{DD} = K$.

Terminology previously used for gems can be extended to jewels in a natural way and will be employed without further explanation. For example, a bigon in a jewel is a circuit that meets only two of the four colours.

It is evident from the definitions that if (X, π, φ, ι) is an imbedding corresponding to K, then $(X, \pi, \varphi, -\iota)$ corresponds to K^S. Hence $K^{SS} = K$. We say that $(X, \pi, \varphi, -\iota)$ is *skew* to (X, π, φ, ι), and the faces of $(X, \pi, \varphi, -\iota)$ are the *zigzag walks* of (X, π, φ, ι). It is clear from the above discussion that in the imbeddings (X, π, φ, ι) and $(X, \pi, \varphi, -\iota)$ the faces and zigzag walks are interchanged. The zigzag walks correspond to the blue-yellow bigons of the jewel (L, Z, F, V) and therefore to the green-yellow bigons of (L, V, F, Z).

Next, consider the gem K^D. In Chapter 2, we established a bijection between the set of red-yellow bigons of K and the set of vertices of a corresponding imbedding I, and another between the set of blue-yellow bigons of K and the set of faces of I. Consequently, in the imbeddings corresponding to K and K^D, the roles of faces and vertices are interchanged. Any imbedding corresponding to K^D is said to be *dual* to one corresponding to K. Note that an imbedding dual to a planar one is also planar.

An imbedding corresponding to K^{DSD} is *phial* to one corresponding to K. In these imbeddings, vertices and zigzag walks are interchanged.

For any imbedding I, imbeddings skew, dual, and phial to I will be denoted by I^S, I^D, and I^{DSD}, respectively. Edges $e \in EI$ and $e^D \in EI^D$ are said to *correspond* if they both correspond to the same bisquare in the gem that encodes I.

Before we present an example of an imbedding and its skew, dual, and phial imbeddings, let us consider the question of whether an edge e should be barred in the dual of an imbedding I. Let $I = (X, \pi, \varphi, \iota)$, $I^D = (X^D, \pi^D, \varphi^D, \iota^D)$, and $e = \{x, \pi x\}$. Let $P_1, P_1^-, P_2, P_2^- \cdots, P_n, P_n^-$ be the orbits of I, and let $(x, 1) \in P_i$ for some $i \in J_n$. For some $j \in J_n$, we have either $(x, -1) \in P_j$ or $(x, -1) \in P_j^-$. We let the corresponding edge e^D be barred in I^D in the former case and unbarred in the latter. Thus, the decision as to whether e^D should be barred in I^D depends on the choice of $P_1, P_2, \ldots, P_n$. (Intuitively speaking, one chooses an arbitrary orientation for each face and then places a bar on e^D if the chosen orientations on the faces to which e^D belongs agree on e^D.) However, the imbeddings resulting from distinct choices are easily seen to be equivalent. Indeed, let $f_1, f_2, \ldots, f_n$ be the faces determined by $P_1, P_2, \ldots, P_n$, respectively. Let $S \subseteq J_n$, and suppose that P_i is replaced by P_i^- for each $i \in S$. Let T be the set of faces f_j for which $j \in S$. Then $\iota^D(e^D)$ is replaced by $-\iota^D(e^D)$ if and only if e belongs to ∂T. Since $f_1, f_2, \ldots, f_n$ correspond to vertices in I^D, it follows immediately that the new imbedding is equivalent to I^D. Note, however, that for a given imbedding I its dual imbedding I^D is not uniquely defined.

Note that if e is monofacial, then e^D is barred if and only if e is consistent in I. This conclusion agrees with what one might expect upon consideration of Example 4.2 (p. 55) with the colours red and blue interchanged.

Example 8.1 *In Figure 8.2 we give an example of an imbedding and its skew, dual, and phial imbeddings. Here* $EI = \{e_1, e_2, e_3, e_4, e_5\}$. *It is evident that the imbedding in Figure 8.2b is indeed* I^S. *Observe next that I has three faces, determined by oriented boundaries* $[e_2]$, $[e_1e_4e_3]$, *and* $[e_3e_5e_4e_5e_1e_2]$, *respectively.*

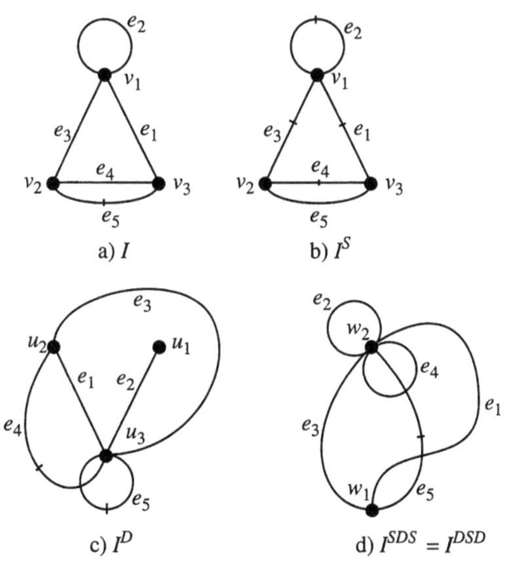

FIGURE 8.2.

These faces are the vertices u_1, u_2, and u_3, respectively, in Figure 8.2c. It is now easily checked that the imbedding in Figure 8.2c is a dual of I. Likewise, I^S has two faces, determined by oriented boundaries $[e_1e_5e_3]$ and $[e_2e_2e_1e_4e_5e_4e_3]$, respectively. These are the vertices w_1 and w_2, respectively, in Figure 8.2d. The imbedding in Figure 8.2d is now easily seen to be the skew of a dual of I^S and hence a phial of I.

Let e be an edge of a graph G, and let K be a gem encoding an imbedding I of G. The gem K^D dual to K encodes an imbedding of a graph. We say that this graph is the *dual* of G with respect to I. The dual of G with respect to an imbedding I will be denoted by $G^{D/I}$, or simply by G^D if the imbedding is understood. Recall that e corresponds to a bisquare of K, which is also a bisquare of K^D and hence an edge e^D of G^D. We say that e *dualises* to e^D (with respect to I). If this is the case, then e^D also dualises to e (with respect to I^D), and we may say simply that e and e^D are *duals* (of each other). If $S \subseteq EG$, then we also say that S *dualises* to the set S^D of duals of the members of S. Thus S and S^D dualise to each other, and we may describe them as *duals* (of each other). If $\mathcal{S}$ is a collection of subsets of EG, then we denote by $\mathcal{S}^D$ the set of their duals.

If G is connected, then K is connected. Hence K^D and G^D are connected.

8.2 Constructing orthogonal graphs

If G is a planar graph, then by Whitney's theorem (p. 95) there exists another planar graph orthogonal to G. Our next theorem uses the concept of duality to construct such a graph.

Theorem 8.1 *Let G be a planar graph. Then the dual of G with respect to a planar imbedding of G is isomorphic to a graph orthogonal to G, and the edge isomorphism is given by duality.*

Proof. For any planar imbedding of G, the set of face boundaries spans $\mathcal{C}(G)$, by Theorem 5.4 (p. 69). The faces of the imbedding are in 1:1 correspondence with the blue-yellow bigons in the gem K that encodes the imbedding. Therefore, they are also in 1:1 correspondence with the red-yellow bigons in K^D and hence with the vertex cocycles of the graph G^D of K^D. Moreover, the edges of G are in 1:1 correspondence with the bisquares of K, that are also the bisquares of K^D and hence the edges of G^D. This correspondence is such that an edge belonging to a given face boundary corresponds to a bisquare of K which meets the blue-yellow bigon corresponding to the face boundary. In other words, the dual of an edge belonging to a given face boundary is a member of the corresponding vertex cocycle of G^D. If we use duality to identify EG with EG^D, then we find that $\mathcal{C}(G) = \mathcal{C}^{\perp}(G^D)$, as required. ■

It follows from Theorem 8.1 that $\mathcal{C}(G^D) = \mathcal{C}^{\perp}(G)$ for a planar graph G.

One observation to be made from the proof of this theorem is that, for any imbedding of a graph, the dual of a face boundary is a vertex cocycle. Similarly, the dual of a vertex cocycle is a face boundary (where the imbedding of the dual graph is taken to be a dual of that of the given graph).

Theorem 8.1 itself has the following converse.

Theorem 8.2 *Orthogonal 2-connected graphs have planar imbeddings encoded by dual gems.*

Proof. Note that orthogonal 2-connected graphs must have no loops, since the existence of a loop in one implies the existence of an isthmus in the other.

Let G and G' be orthogonal 2-connected graphs, and choose $v \in VG'$. Then ∂v is a vertex cocycle A in G' and hence a bond because otherwise $G' - \{v\}$ would not be connected. A must be a cycle in G by orthogonality, and therefore a circuit since A is a bond in G'. Since the vertex cocycles of G' span $\mathcal{C}^{\perp}(G')$, they are circuits of G that span $\mathcal{C}(G)$. In fact, they constitute a spanning circuit double cover of G, since each edge of G' belongs to just two vertex cocycles. By Corollary 6.16, G has a planar imbedding for which each component of the spanning circuit double cover constitutes a face boundary.

Let H be the gem encoding this imbedding. Its blue-yellow bigons therefore correspond to the circuits of G that are vertex cocycles in G'. Then H^D encodes an imbedding of a graph G^D, and this imbedding is again planar. Moreover, G' and G^D are isomorphic: The vertex isomorphism from VG' onto VG^D maps a vertex v of G' onto a red-yellow bigon of H^D that is a blue-yellow bigon of H corresponding to the circuit ∂v of G, and the edge isomorphism from EG' onto EG^D maps an edge of $EG' = EG$ onto the corresponding red-blue bigon of H. ■

8.3 Duality for planar graphs

We now establish some properties of duality for planar graphs and show how they generalise. The treatment follows Richter and Shank (1984).

Theorem 8.3 *Let G^D be the dual of a planar graph G with respect to a planar imbedding of G, and let $C \subseteq EG$.*

(1) *C is a cycle of G if and only if C^D is a cocycle of G^D.*

(2) *C is a cocycle of G if and only if C^D is a cycle of G^D.*

Proof. Theorem 8.1 shows that each member of $\mathcal{C}(G)$ dualises to a member of $\mathcal{C}^{\perp}(G^D)$, and vice versa. ■

It follows immediately from Theorem 8.3 that, for planar graphs, circuits and bonds are duals of each other.

A generalisation of Theorem 8.3 has been proved by Edmonds (1965). Let I be an imbedding of a graph G. We have already noted in Chapter 2 that the boundary of each face is a cycle. The subspace of $\mathcal{C}(G)$ spanned by such cycles is called the *boundary space* of G (with respect to I) and its members *boundaries*. This space will be denoted by $\mathcal{F}(G; I)$ or by $\mathcal{F}(G)$ if the reference to I is otherwise clear. If I is planar, then the boundary space of G is $\mathcal{C}(G)$ by Theorem 5.4; but if I is non-planar then G has cycles that are not boundaries. However, we have the following generalisation of Theorem 8.3.

Theorem 8.4 *Let I be an imbedding of a graph G, and let $C \subseteq EG$.*

(1) *C is a boundary of G with respect to I if and only if C^D is a cocycle of G^D.*

(2) *C is a cocycle of G if and only if C^D is a boundary of G^D with respect to I^D.*

Proof. (1) C is a boundary if and only if it is of the form $\sum \mathcal{B}$ for some set $\mathcal{B}$ of face boundaries. But $C^D = (\sum \mathcal{B})^D = \sum \mathcal{B}^D$, which is a sum of vertex cocycles, and hence a cocycle of G^D. Conversely, any cocycle of G^D is a sum of vertex cocycles and hence the dual of a sum of face boundaries.

The proof of (2) is similar. ■

Corollary 8.5 *Let I be an imbedding of a connected graph G. Then*

$$\dim \mathcal{F}(G) = |FI| - 1.$$

Proof. From Theorem 8.4(1), we have

$$\begin{aligned} \dim \mathcal{F}(G) &= \dim \mathcal{C}^{\perp}(G^D) \\ &= |VG^D| - 1 \\ &= |FI| - 1. \blacksquare \end{aligned}$$

For planar imbeddings we also have the following result.

Theorem 8.6 *Let I be a planar imbedding of a planar connected graph G. Let T be the edge set of a spanning tree of G. Then $G^D - T^D$ is a spanning tree of G^D.*

Proof. Since $G[T]$ is connected, T meets every bond of G. Therefore T^D meets every circuit of G^D, so that $G^D - T^D$ is a forest. Moreover T includes no circuit, and so T^D includes no bond. In particular $EG^D - T^D$ must meet every vertex cocycle, and $G^D - T^D$ is therefore spanning.

Finally, suppose that $G^D - T^D$ is not connected. Since G is connected, so is G^D. Therefore T^D includes a non-empty cocycle of G^D. Thus, we obtain the contradiction that T^D includes a bond. ■

If I is non-planar, however, only the following weaker conclusion holds.

Theorem 8.7 *Let I be any imbedding of a connected graph G. Let T be the edge set of a spanning tree of G. Then the graph $G^D - T^D$ is connected.*

Proof. Suppose not. Again T^D includes a non-empty cocycle of G^D, since G^D is connected. Therefore T includes a non-empty boundary of G by Theorem 8.4, and hence a non-empty cycle. We thus have the contradiction that T includes a circuit. ■

One corollary is the following result of Biggs (1976).

Corollary 8.8 *Let I be an orientable imbedding of a connected graph G, and let T be the edge set of a spanning tree of G. Then there exists a spanning tree of $G^D - T^D$ whose edge set consists of all but $2g$ edges of $EG^D - T^D$, where g is the genus of I.*

Proof. Let U^D be the edge set of a spanning tree of $G^D - T^D$. Since

$$|VG| - |EG| + |VG^D| = 2 - 2g,$$

we have

$$\begin{aligned} |U^D| &= |VG^D| - 1 \\ &= 1 + |EG| - |VG| - 2g. \end{aligned}$$

Hence

$$\begin{aligned} |EG^D| - |T^D| - |U^D| &= |EG| - |T| - |EG| + |VG| - 1 + 2g \\ &= 2g, \end{aligned}$$

since $|T| = |VG| - 1$. ■

In the corresponding result for non-orientable imbeddings, of course $2g$ is replaced by the cross-cap number of the imbedding.

A related theorem is the following observation of Mohar (1987).

Theorem 8.9 *Let I be an orientable imbedding of a connected graph G. Then* $\dim \mathcal{F}^{\perp}(G) = 2g$, *where g is the genus of I.*

Proof. We have

$$\begin{aligned} \dim \mathcal{F}^{\perp}(G) &= \dim \mathcal{C}(G) - \dim \mathcal{F}(G) \\ &= |EG| - |VG| + 1 - (|FI| - 1) \\ &= |EG| - |VG| + 1 - (|EG| - |VG| + 1 - 2g) \\ &= 2g. \ ■ \end{aligned}$$

Similarly, if I were replaced by a non-orientable imbedding of cross-cap number k, then we would have $\dim \mathcal{F}^{\perp}(G) = k$.

By Theorem 5.4, the face boundaries of a planar imbedding of a planar graph G span $\mathcal{C}(G)$. In fact, if G is 2-connected, then Corollary 1.36 shows that a basis of circuits for $\mathcal{C}(G)$ may be constructed by discarding an arbitrary component of a spanning circuit double cover, since any circuit of the double cover is the sum of the others. The double cover may be obtained from the face boundaries of the planar imbedding. For non-planar graphs, however, this procedure yields an insufficient number of circuits to give a basis of $\mathcal{C}(G)$. We proceed to show how to supplement the set of circuits so obtained. First, we derive a lemma.

Lemma 8.10 *Let I be an imbedding of a connected graph G, and let T be the edge set of a spanning tree of G. Let U be a subset of $EG - T$ such that U^D is the edge set of a spanning tree of $G^D - T^D$. Let X be a non-empty subset of $EG - (T \cup U)$. Then the sum of the fundamental circuits of G (with respect to T) that meet X is not a boundary.*

Proof. Let B be any non-empty boundary. Then B^D is a cocycle of G^D and therefore meets U^D. Hence $B \cap U \neq \emptyset$, and the result follows immediately. ∎

Theorem 8.11 *Let I be an imbedding of a connected graph G, and let T be the edge set of a spanning tree of G. Let $\mathcal{R}$ be a set consisting of all but one of the face boundaries of G. Let $U \subseteq EG - T$, and let $\mathcal{S}$ be the set of fundamental circuits of G that do not meet U. Then U^D is the edge set of a spanning tree of G^D if and only if $\mathcal{R} \cap \mathcal{S} = \emptyset$ and $\mathcal{R} \cup \mathcal{S}$ is a basis of $\mathcal{C}(G)$.*

Proof. Recall from Corollary 8.5 that $\mathcal{R}$ is a basis of $\mathcal{F}(G)$.
Suppose that $\mathcal{R} \cap \mathcal{S} = \emptyset$ and $\mathcal{R} \cup \mathcal{S}$ is a basis of $\mathcal{C}(G)$. Thus

$$\begin{aligned} |\mathcal{R}| + |\mathcal{S}| &= |\mathcal{R} \cup \mathcal{S}| \\ &= |EG| - |VG| + 1. \end{aligned}$$

But

$$\begin{aligned} |\mathcal{S}| + |U| &= |EG| - |T| \\ &= |EG| - |VG| + 1 \\ &= |\mathcal{R}| + |\mathcal{S}|. \end{aligned}$$

Hence $|U| = |\mathcal{R}| = |VG^D| - 1$.

By Theorem 1.18 it therefore suffices to show that $G^D[U^D]$ is a connected spanning subgraph of G^D. Suppose not. Then there exists a non-empty subset X of $EG - U$ such that X^D is a cocycle of G^D. Since X is a boundary by Theorem 8.4, the cycle X is a sum of members of $\mathcal{R}$, and $X - T \neq \emptyset$. Let Y be the sum of the members of $\mathcal{S}$ that meet $X - T$, and let $Z = X + Y$. Since $\mathcal{R} \cup \mathcal{S}$ is a basis of $\mathcal{C}(G)$ and $X \neq \emptyset$, it follows that $Z \neq \emptyset$. But $Y \cap (EG - T) = X \cap (EG - T)$, so that $Z \subseteq T$, in contradiction to the fact that Z is a non-empty cycle.

Conversely, suppose that $G^D[U^D]$ is a spanning tree of G^D. Then $|U^D| = |VG^D| - 1 = |\mathcal{R}|$. Thus

$$\begin{aligned} |\mathcal{R}| + |\mathcal{S}| &= |\mathcal{R}| + |EG| - |T| - |U^D| \\ &= |EG| - |VG| + 1 \\ &= \dim \mathcal{C}(G). \end{aligned}$$

Hence, it suffices to establish the linear independence of the members of $\mathcal{R}$ and $\mathcal{S}$.

Choose $\mathcal{P} \subseteq \mathcal{R}$ and $\mathcal{Q} \subseteq \mathcal{S}$ such that $\sum \mathcal{P} + \sum \mathcal{Q} = \emptyset$. Then $\sum \mathcal{Q} = \sum \mathcal{P}$, so that $\sum \mathcal{Q}$ is a boundary. Lemma 8.10 therefore implies that $\mathcal{Q} = \emptyset$, and so $\sum \mathcal{P} = \emptyset$. Since $\mathcal{R}$ is a basis of $\mathcal{F}(G)$, it follows that $\mathcal{P} = \emptyset$, as required. ∎

Example 8.2 *Let the vertex set of* $K_{4,4}$ *be* $\{v_1, v_2, v_3, v_4, w_1, w_2, w_3, w_4\}$, *where* v_i *is joined to* w_j *by an edge* e_{ij} *for each* $(i, j) \in J_4 \times J_4$. *It is easy to find an imbedding of* $K_{4,4}$ *for which the face boundaries are* $F_{12,14} = \{e_{11}, e_{14}, e_{21}, e_{24}\}$, $F_{12,23} = \{e_{12}, e_{13}, e_{22}, e_{23}\}$, $F_{23,12} = \{e_{21}, e_{22}, e_{31}, e_{32}\}$, $F_{23,34} = \{e_{23}, e_{24}, e_{33}, e_{34}\}$, $F_{34,14} = \{e_{31}, e_{34}, e_{41}, e_{44}\}$, $F_{34,23} = \{e_{32}, e_{33}, e_{42}, e_{43}\}$, $F_{14,12} = \{e_{11}, e_{12}, e_{41}, e_{42}\}$, *and* $F_{14,34} = \{e_{13}, e_{14}, e_{43}, e_{44}\}$. *Let* T *be the spanning tree with edge set*

$$\{e_{11}, e_{12}, e_{13}, e_{14}, e_{21}, e_{31}, e_{41}\}.$$

Then the subset

$$U = \{e_{24}, e_{33}, e_{42}, e_{23}, e_{34}, e_{32}, e_{43}\}$$

of $EK_{4,4} - ET$ *has the property that* U^D *is the edge set of a spanning tree of* $K_{4,4}^D$. *The only fundamental circuits of* G *that do not meet* U *are* $A = \{e_{22}, e_{21}, e_{11}, e_{12}\}$ *and* $B = \{e_{44}, e_{41}, e_{11}, e_{14}\}$. *It follows from Theorem 8.11 that*

$$\{A, B, F_{12,14}, F_{12,23}, F_{23,12}, F_{23,34}, F_{34,14}, F_{34,23}, F_{14,12}\}$$

is a basis of $\mathcal{C}(K_{4,4})$.

8.4 The zigzag space

We now turn our attention to the zigzag walks of a graph G with a given imbedding I. By definition, each is a face of the imbedding I^S of G, and hence its boundary is a cycle. Such boundaries are called *zigzag boundaries*. We call the subspace of $\mathcal{C}(G)$ spanned by them the *zigzag space* of G. It is denoted by $\mathcal{Z}(G; I)$, or simply by $\mathcal{Z}(G)$ if the reference to I is otherwise clear.

Let (H, V, F) be the gem encoding the imbedding I, and let (H, V, F, Z) be the associated jewel J. Let P be a zigzag walk of G. It is represented by a green-yellow bigon of J. It dualises to a green-yellow bigon of the jewel (H, F, V, Z), and this bigon represents a zigzag walk in G^D. Of course, the boundary of this zigzag walk is a cycle of G^D.

To the union of the green-yellow bigons of J let us now assign a bijective orientation σ. If a and b are the two green edges joining vertices of a given red-blue bisquare Y, two possibilities emerge: Either $\sigma(a)$ and $\sigma(b)$ are ends of the same red edge or they are not. We say that the edge e of G corresponding to Y is *skew-consistent* in I (with respect to σ) in the former case and *skew-coherent* in I in the latter. Observe that if e is monofacial in I^S, then these properties are independent of the choice of σ. Moreover, in this case e is skew-consistent (skew-coherent) in I if and only if it is consistent (coherent) in I^S. If e is skew-consistent, then we also say that e is oriented *to* the end corresponding to the red-yellow bigon that passes through $\sigma(a)$ and $\sigma(b)$, and *from* its other end. (A skew-consistent loop is thus oriented both to and from its unique end.) The bijective nature of σ implies that for any vertex v of G there are exactly as many edges oriented to v as from. It follows immediately that the set of skew-consistent edges constitutes a cycle of G. An inspection of $J[VY]$ also reveals immediately that e is skew-consistent if and only if e^D is skew-coherent. If we apply this observation in the case where G is planar, we infer that there exists a pseudo-partition of EG, of cardinality 2, whose elements are a cycle and a cocycle, respectively. Later we shall extend this conclusion to any graph G.

Example 8.3 *Figure 8.3 [which appears in Rosenstiehl and Read (1978)] represents a (planar) imbedding I of a graph G. The sets*

$$[e_1e_7e_8e_3e_4e_{15}e_{12}e_{11}e_5e_1e_2e_8e_9e_{16}e_{13}e_{12}e_{10}e_5e_6e_{14}e_{16}e_4],$$
$$[e_7e_2e_3e_9e_{14}e_{13}e_{15}e_{10}e_{11}e_6]$$

are oriented boundaries of faces of I^S. Let J be the jewel associated with the gem that encodes I. There exists a bijective orientation of the union of the green-yellow bigons of J with respect to which the edges e_5, e_7, e_8, e_9, e_{10}, e_{15}, e_{16} are skew-consistent and the others skew-coherent. The skew-consistent edges constitute a cycle, and the set of skew-coherent edges is the cocycle $\partial\{v_1, v_6\}$.

We now investigate more closely those edges that are monofacial in the imbedding I^S. They belong to just one zigzag walk. Let e be such an edge, and let W be the unique zigzag walk to which it belongs. There are two boundary segments of W with respect to e. Each of these will be said to be *determined* by e. For instance, the edge e_5 in Example 8.3 determines the boundary segments

$$\{e_1, e_2, e_8, e_9, e_{16}, e_{13}, e_{12}, e_{10}\}, \{e_6, e_{14}, e_{16}, e_1, e_7, e_8, e_3, e_{15}, e_{12}, e_{11}\}.$$

The following theorem is immediate from the definitions.

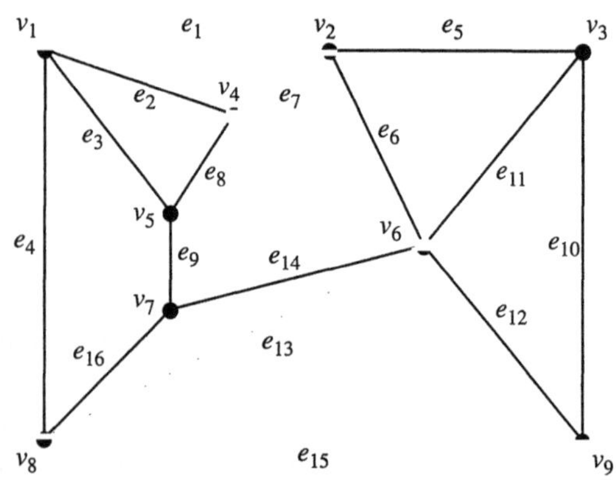

FIGURE 8.3.

Theorem 8.12 *Let G be a graph with imbedding I. Let e be monofacial in I^S, and let T be a boundary segment determined by e.*

(1) *If e is skew-consistent, then $T \cup \{e\}$ is a cycle.*

(2) *If e is skew-coherent, then T is a cycle.*

In view of Theorem 8.12, we say that a skew-consistent monofacial edge is *cyclic* in I.

Dually, if e is skew-consistent, then $T^D = T$ is a cycle of G^D; and if e is skew-coherent, then $T^D \cup \{e^D\} = T \cup \{e\}$ is a cycle of G^D. A skew-coherent monofacial edge is said to be *cocyclic* in I.

Theorem 8.13 *Let G be a graph with imbedding I. For each $e \in EG$ the number of the following statements that hold is odd:*

(1) *there is a cycle A such that $e \in A$ and $A^D + \{e^D\} \in \mathcal{C}(G^D)$;*

(2) *there is a cycle B such that $e \notin B$ and $B^D + \{e^D\} \in \mathcal{C}(G^D)$;*

(3) *there is a cycle C such that $e \in C$ and $C^D \in \mathcal{C}(G^D)$.*

Proof. If e is monofacial in I^S, then, by Theorem 8.12 and its dual, (1) holds if e is skew-consistent and (2) holds otherwise. If e is bifacial in I^S, then e belongs to a zigzag boundary, which necessarily satisfies (3).

Suppose now that (1) and (2) both hold. Thus there are cycles A and B such that $A^D + \{e^D\}$ and $B^D + \{e^D\}$ are both cycles of G^D and $e \in A - B$. Let $C = A + B$. Then C is a cycle and $e \in C$. Moreover,

$$\begin{aligned} C^D &= A^D + B^D \\ &= (A^D + \{e^D\}) + (B^D + \{e^D\}), \end{aligned}$$

which is a cycle of G^D. Thus (3) holds.

Suppose next that (3) holds. Thus there is a cycle C such that $e \in C$ and C^D is a cycle of G^D. If (1) holds, then there is also a cycle A such that $e \in A$ and $A^D + \{e^D\}$ is a cycle of G^D. Let $B = A + C$. Then B is a cycle such that $e \notin B$, and $B^D + \{e^D\} \in \mathcal{C}(G^D)$. Hence (2) holds. Similarly, if (2) holds then so does (1). ■

Corollary 8.14 *Let I be a planar imbedding of a planar graph G, and let $e \in EG$. Then exactly one of the following statements holds:*

(1) *there is a cycle A such that $e \in A$ and $A + \{e\} \in \mathcal{C}^\perp(G)$;*

(2) *there is a cycle B such that $e \notin B$ and $B + \{e\} \in \mathcal{C}^\perp(G)$;*

(3) *there is a cycle C such that $e \in C$ and $C \in \mathcal{C}^\perp(G)$.*

Proof. By Theorems 8.1 and 8.13, it suffices to obtain a contradiction if (1)–(3) all hold. However, we need only the assumption that (2) and (3) hold. Then $\{e\}$ is the sum of the cycle $B+C$ and the cocycle $B+C+\{e\}$; hence $\{e\} \in \mathcal{C}(G)+\mathcal{C}^{\perp}(G)$. For any cocycle Y, we see that $|C \cap Y|$ is even, since $C \in \mathcal{C}(G)$. Similarly, $|C \cap Z|$ is even for any cycle Z. Hence $C \in (\mathcal{C}(G)+\mathcal{C}^{\perp}(G))^{\perp}$. It follows in particular that $|C \cap \{e\}|$ is even, in contradiction to the assumption that $e \in C$. ■

Example 8.4 *The edge e_5 in Example 8.3 belongs to the cycle*

$$A = \{e_5, e_1, e_2, e_8, e_9, e_{16}, e_{13}, e_{12}, e_{10}\}.$$

$A + \{e_5\}$ is the cocycle $\partial\{v_1, v_5, v_8, v_9\}$. The edge e_1 belongs to the cocycle

$$B + \{e_1\} = \partial\{v_1, v_3, v_4, v_9\},$$

where B is the cycle $\{e_7, e_8, e_3, e_4, e_{15}, e_{12}, e_{11}, e_5\}$. The edge e_2 belongs to the cycle

$$C = \{e_2, e_3, e_9, e_{14}, e_{13}, e_{15}, e_{10}, e_{11}, e_6, e_7\} = \partial\{v_4, v_5, v_6, v_9\}.$$

8.5 The principal edge tripartition for planar graphs

For a planar graph G, Corollary 8.14 yields a pseudo-partition of EG called the *principal edge tripartition* of G. Later, we shall drop the restriction of planarity. In the general case, and in the remainder of this chapter, we denote the principal edge tripartition of G by (P, Q, R), where P,Q, and R are the sets of edges satisfying (1), (2), and (3), respectively, of Corollary 8.14.

We proceed to develop an easy method for finding P, Q, and R in the case of a planar graph. Several preliminary results are required. First, let us define two distinct edges of a jewel to be *conjugate* if they both join two vertices of the same red-blue bigon and have the same colour.

Lemma 8.15 *Let I be a planar imbedding of a planar graph G.*

(1) *If G is Eulerian, then any edge monofacial in I^S is cyclic in I.*

(2) *If G is bipartite, then any edge monofacial in I^S is cocyclic in I.*

Proof. If G is Eulerian, every face boundary of G^D is of even cardinality. Since I^D is also planar, these face boundaries span $\mathcal{C}(G^D)$ by Theorem 5.4 (p. 69). It follows that G^D is bipartite by Theorem 1.22 (p. 12). Let J^D be the jewel associated with the gem K^D that encodes the imbedding I^D, and let σ be a bijective orientation of the set of green and yellow edges of J^D. Since I^D is planar, K^D is bipartite.

Let e be an edge of G that is monofacial in I^S, and let B be the corresponding bisquare of K^D. B is also a red-blue bigon of the jewel J^D. There are two conjugate green edges, a_0 and d, of J^D that join vertices of VB; and since e is monofacial

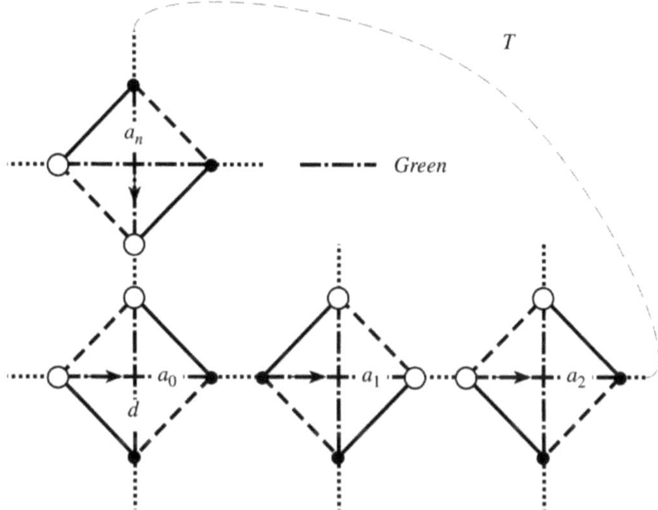

FIGURE 8.4.

in I^S, it follows that a_0 and d belong to a common green-yellow bigon X. Thus $X - \{a_0, d\}$ is the union of two disjoint paths. Of these two paths, let T be the one that has $\sigma(a_0)$ as an end, and let $\{a_1, a_2, \ldots, a_n\}$ be the set of green edges of T.

Suppose that e is cocyclic in I. Then e^D is skew-consistent in I^D. Therefore, $\sigma(a_0)$ and $\sigma(d)$ are ends of the same red edge. (See Figure 8.4.) Since G^D is bipartite, its vertices may be coloured black or white so that each edge joins vertices of different colours. The colour of each vertex of G^D may be assigned to the vertices of the corresponding red-yellow bigon of K^D. Each green edge therefore joins vertices of different colours, whereas each yellow edge joins similarly coloured vertices. As the ends of the path $\{a_0\} \cup T$ have the same colour, we infer that n is odd.

The ends of T are joined by a blue edge b_0. For each $i \in J_n$, let b_i and c_i be blue and red neighbours, respectively, of a_i such that $\{a_i, b_i, c_i\}$ is a triangle Y_i. Then $\{b_0\} \cup (T + \bigcup_{i=1}^{n} Y_i)$ is a cycle C in K^D (since it has no green edges). But $|C| = 3n + 2 - 2m$, where m is the number of bisquares whose vertices are all in VT. Thus $|C|$ is odd. This result contradicts the fact that K^D must be bipartite.

We conclude that e is cyclic in I. The proof of (1) is now complete, and (2) follows by duality. ■

Lemma 8.16 *Each zigzag boundary in a planar graph is both a cycle and a cocycle.*

Proof. We have already observed that each zigzag boundary is a cycle.

Let J be a jewel associated with a gem K encoding a planar imbedding I of a graph G. Let X be a green-yellow bigon in J corresponding to a zigzag walk of I. X is also a green-yellow bigon in the jewel J^D associated with K^D. It therefore corresponds to a zigzag walk of I^D whose boundary is a cycle of G^D. The result follows by Theorem 8.3. ■

Lemma 8.17 *Let I be any imbedding of a connected graph G. Then any non-empty proper subset of the set of all face boundaries of I is linearly independent.*

Proof. Let K be the gem encoding I. Let $\mathcal{S}$ be a non-empty set of blue-yellow bigons of K such that $\bigcup \mathcal{S}$ contains an even number of blue edges of each bisquare. It suffices to show that every blue-yellow bigon must be a member of $\mathcal{S}$. Choose $C \in \mathcal{S}$ and $e \in \beta(C)$, and let A and B be the red-yellow bigons passing through the ends of e. Note that the blue edges adjacent to the yellow neighbours of e must be in C, and that both blue edges of the bisquare containing e are in $\bigcup \mathcal{S}$ by assumption. Proceeding inductively, we find that $\partial VA \cup \partial VB \subseteq \bigcup \mathcal{S}$. Since K is connected, a further induction then shows that every blue edge of K is in $\bigcup \mathcal{S}$, as required. ■

The following corollary is immediate.

Corollary 8.18 *Let I be an imbedding of a connected graph G. If I has more than one face, then some edge is bifacial in I.*

Theorem 8.19 *Let I be any planar imbedding of a planar connected graph G. Then I has a unique zigzag walk if and only if no non-empty cycle is also a cocycle.*

Proof. Suppose that I has more than one zigzag walk. Then I^S has a bifacial edge by Corollary 8.18. Hence I has a non-empty zigzag boundary, which is a member of $\mathcal{C}(G) \cap \mathcal{C}^{\perp}(G)$ by Lemma 8.16.

On the other hand, suppose that I has a unique zigzag walk, and that there exists a non-empty member D of $\mathcal{C}(G) \cap \mathcal{C}^{\perp}(G)$. Recall that every edge is cyclic or cocyclic in I, and note that D is both a cycle and a cocycle of the graphs $G - \{e\}$ and G_e for any edge $e \in EG - D$. Either the former graph or the latter has a planar imbedding with a unique zigzag walk, depending on whether e is cyclic or cocyclic in I. Repetition of this argument inductively yields a graph H, with edge set D and a planar imbedding with a unique zigzag walk, such that D is both a cycle and a cocycle of H. Hence H is Eulerian and bipartite. By Lemma 8.15, we now obtain the contradiction that any edge of H is both cyclic and cocyclic in I. ■

Theorem 8.20 *Let I be a planar imbedding of a planar connected graph G. Suppose that the number of zigzag walks of I is k. Then any $k-1$ zigzag boundaries constitute a basis of $\mathcal{C}(G) \cap \mathcal{C}^{\perp}(G)$.*

Proof. For $k = 1$ the result holds by Theorem 8.19. Assume as an inductive hypothesis that it holds whenever the number of zigzag walks is less than k, where $k > 1$. Let $\mathcal{Z}$ be the set of zigzag boundaries of I, and choose $Z \in \mathcal{Z}$. Since $k > 1$, it follows from Lemma 8.17 that every zigzag boundary must be non-empty. Therefore, some edge e belongs to Z and to another zigzag boundary Y. It follows that $(\mathcal{Z} - \{Z, Y\}) \cup \{Z + Y\}$ is the set of zigzag boundaries of an imbedding of $G-\{e\}$. By the inductive hypothesis, $\mathcal{Z}-\{Z, Y\}$ is therefore a basis for $\mathcal{C}(G-\{e\}) \cap \mathcal{C}^{\perp}(G-\{e\})$.

Now choose $D \in \mathcal{C}(G) \cap \mathcal{C}^{\perp}(G)$. If $e \notin D$, then D is both a cycle and a cocycle of $G-\{e\}$ and hence a sum of members of $\mathcal{Z}-\{Z, Y\}$. Suppose that $e \in D$. By Lemma 8.16, Y is both a cycle and a cocycle, and therefore $D+Y$ is both a cycle and a cocycle of $G-\{e\}$. Thus $D+Y$ is a sum of members of $\mathcal{Z}-\{Z, Y\}$, and so D is a sum of members of $\mathcal{Z}-\{Z\}$. Hence $\mathcal{Z}-\{Z\}$ spans $\mathcal{C}(G) \cap \mathcal{C}^{\perp}(G)$. Since $k>1$, linear independence follows from Lemma 8.17. ∎

Corollary 8.21 *If G is a planar graph with a planar imbedding I, then*

$$\mathcal{Z}(G)=\mathcal{C}(G) \cap \mathcal{C}^{\perp}(G).$$

For any graph G, an element of $\mathcal{C}(G) \cap \mathcal{C}^{\perp}(G)$ is called a *bicycle*. Each edge in a bicycle is said to be *bicyclic*.

Theorem 8.22 *Let I be a planar imbedding of a planar graph G.*

(1) *If T is a boundary segment in I^S determined by an edge e that is cyclic in I, then $T \cup \{e\}$ is a cycle and T a cocycle.*

(2) *If T is a boundary segment in I^S determined by an edge e that is cocyclic in I, then $T \cup \{e\}$ is a cocycle and T a cycle.*

(3) *The boundary of each zigzag walk is a bicycle.*

Proof. Let e be an edge that is cyclic in I. That $T \cup \{e\}$ is a cycle follows from Theorem 8.12. Since e is cocyclic in I^D, T is a cycle in G^D, and hence a cocycle in G by duality. The proof of (1) is now complete, and (2) follows similarly. Part (3) follows immediately from Corollary 8.21. ∎

The following corollary, which is immediate, shows how to find the principal edge tripartition of a planar graph.

Corollary 8.23 *Let I be a planar imbedding of a planar graph G with principal edge tripartition (P, Q, R). Then*

(1) *P is the set of edges that are both monofacial and consistent in I^S;*

(2) *Q is the set of edges that are monofacial and coherent in I^S;*

(3) *R is the set of edges that are bifacial in I^S.*

Thus P, Q and R are the sets of cyclic, cocyclic, and bicyclic edges, respectively.

Example 8.5 *Figure 8.5 shows the principal edge tripartition of the graph in Example 8.3.*

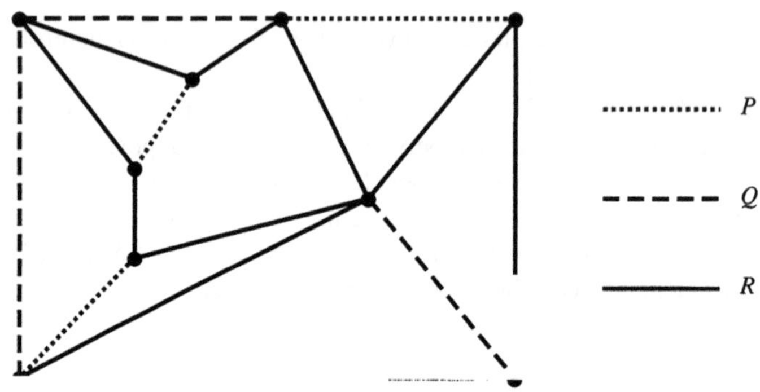

FIGURE 8.5.

8.6 Walks

Suppose that I is a planar imbedding of a planar connected graph G for which $\mathcal{C}(G) \cap \mathcal{C}^{\perp}(G) = \{\emptyset\}$. Then, by Theorem 8.19, I has only one zigzag walk W. Choose $e \in EG$, and let T and U be the boundary segments of W with respect to e. If $e \in P$, then Theorem 8.22(1) shows that $T \cup \{e\}$ is a cycle and T a cocycle. Similarly, if $e \in Q$, then Theorem 8.22(2) shows that T is a cycle and $T \cup \{e\}$ a cocycle.

Now recall that W is the only face of I^S. Let $[e_1 e_2 \cdots e_m]$ be an oriented boundary of W, where $m = 2|EG|$. Associated with this oriented boundary we have a string $e_1 e_2 \cdots e_m$ of edges of G with the following properties:

(1) there is a function σ from J_m onto VG such that e_1 joins $\sigma(m)$ to $\sigma(1)$ and, for each $i \in J_m - \{1\}$, e_i joins $\sigma(i-1)$ to $\sigma(i)$;

(2) for each $e \in EG$ there are exactly two values of $i \in J_m$ for which $e = e_i$;

(3) if $e \in P$ and $e = e_i = e_j$, where $i < j$, then $\sum_{k=i}^{j-1} \{e_k\}$ is a cycle and $\sum_{k=i+1}^{j-1} \{e_k\}$ a cocycle;

(4) if $e \in Q$ and $e = e_i = e_j$, where $i < j$, then $\sum_{k=i}^{j-1} \{e_k\}$ is a cocycle and $\sum_{k=i+1}^{j-1} \{e_k\}$ a cycle.

[Properties (3) and (4) are illustrated in Figures 8.6a and 8.6b, respectively.]

This observation motivates the following definitions.

A string $e_1 e_2 \ldots e_m$ of edges in a graph G is a *walk* W if there exists a function σ from J_m into VG such that $\sigma(1)$ is an end of e_1 and e_i joins $\sigma(i-1)$ to $\sigma(i)$ for each $i \in J_m - \{1\}$. We call σ a *sense* function for W. (It is not unique in the case where there exist distinct vertices v and w in G such that each edge of W joins v to w.) Any element of the set $[W] \cup [W^{-1}]$ is necessarily also a walk and is said to be *equivalent* to W. Equivalence of walks clearly defines an equivalence relation.

If $e_1 e_2 \cdots e_m$ is a walk W and e is an edge, then the number of values of i for which $e_i = e$ is said to be the number of times that W *uses* e. Thus, a walk uses each of its components at least once.

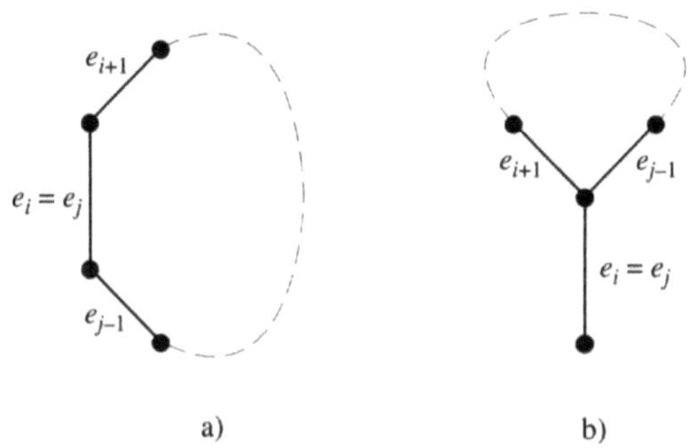

FIGURE 8.6.

The *boundary* of a walk W is the set of edges used an odd number of times and will be denoted by ∇W. If $\mathcal{S}$ is a set of walks, then the *boundary* $\nabla\mathcal{S}$ of $\mathcal{S}$ is defined as the sum of the boundaries of the members of $\mathcal{S}$. We say that W is *closed* if its boundary is a cycle. If W is a closed walk $e_1e_2\cdots e_m$ with sense function σ, it follows that e_1 joins $\sigma(m)$ to $\sigma(1)$.

Now let $e_1e_2\cdots e_m$ be a closed walk W that uses some edge e just twice. Without loss of generality, we may suppose that $e = e_1$. Then there is a unique integer $i > 1$ such that $e = e_i$. It follows that $e_2e_3\cdots e_{i-1}$ and $e_{i+1}e_{i+2}\cdots e_m$ are walks, and we refer to their boundaries as the *boundary segments* of W determined by e_1. Since W is closed, if either boundary segment is a cycle, then so is the other.

If W is a closed walk $e_1e_2\cdots e_m$, then e_1X^{-1} is also a closed walk, where $X = e_2e_3\cdots e_m$. This closed walk is the *conjugate* W^* of W. Clearly, $W^{**} = W$. Moreover, W^* is equivalent to W.

Following Eggleton and Skilton (1984) we call a walk $e_1\cdots e_m$ in G a *double tracing* if it uses each edge of G exactly twice. Thus, every double tracing is closed.

Next, let W be a double tracing in a planar connected graph G, and for any edge $e \in EG$ let the boundary segments of W determined by e be $A(e)$ and $B(e)$, respectively. Suppose that $A(e) \cup \{e\}$ and $A(e)$ are a cycle and a cocycle, respectively, for each $e \in P$, and a cocycle and a cycle, respectively, for each $e \in Q$. Then W is called an *algebraic diagonal* of G. [Note that we must have $A(e) + B(e) = \emptyset$ since W is a double tracing. Hence $A(e) = B(e)$, and so the concept is well defined.]

With these definitions we may now assert that if G is a connected planar graph with no non-empty bicycles then G has an algebraic diagonal since G has a unique zigzag walk by Theorem 8.19. A converse has been proved by Rosenstiehl and Read (1978). In the next section, however, we prove a more general result, in which the condition that $\mathcal{C}(G) \cap \mathcal{C}^{\perp}(G) = \{\emptyset\}$ is dropped.

8.7 Principal cycles and principal cocycles

First, we need to generalise to non-planar graphs some of the concepts that we have introduced. Let us begin with the following theorem.

Theorem 8.24 *For any graph G, we have*

$$(\mathcal{C}(G) \cap \mathcal{C}^{\perp}(G))^{\perp} = \mathcal{C}(G) + \mathcal{C}^{\perp}(G).$$

Proof. The theorem is a special case of a more general result of linear algebra and can be proved directly as follows. Let $Y = W + Z$, where W is a cycle and Z a cocycle, and let X be a bicycle. Since X is a cocycle, $|X \cap W|$ is even. Similarly, $|X \cap Z|$ is even since X is a cycle. As

$$\begin{aligned} X \cap Y &= X \cap (W + Z) \\ &= (X \cap W) + (X \cap Z), \end{aligned}$$

it follows that $|X \cap Y|$ is even. Hence

$$\mathcal{C}(G) + \mathcal{C}^{\perp}(G) \subseteq (\mathcal{C}(G) \cap \mathcal{C}^{\perp}(G))^{\perp},$$

and the theorem follows since

$$\begin{aligned} \dim(\mathcal{C}(G) + \mathcal{C}^{\perp}(G)) &= \dim \mathcal{C}(G) + \dim \mathcal{C}^{\perp}(G) - \dim(\mathcal{C}(G) \cap \mathcal{C}^{\perp}(G)) \\ &= |EG| - \dim(\mathcal{C}(G) \cap \mathcal{C}^{\perp}(G)) \\ &= \dim(\mathcal{C}(G) \cap \mathcal{C}^{\perp}(G))^{\perp}. \blacksquare \end{aligned}$$

Now choose an edge $e \in EG$. Suppose that there is a bicycle B such that $e \in B$. Then $|\{e\} \cap B| = 1$, and it follows from Theorem 8.24 that $\{e\} \notin \mathcal{C}(G) + \mathcal{C}^{\perp}(G)$.

Conversely, if e is not a member of any bicycle, then $|\{e\} \cap B| = 0$ for every bicycle B, and Theorem 8.24 therefore shows that $\{e\} \in \mathcal{C}(G) + \mathcal{C}^{\perp}(G)$. In this case, we may write $\{e\} = C(e) + D(e)$, where $C(e)$ is a cycle and $D(e)$ a cocycle. Let us define a cycle (cocycle) A to be a *principal* cycle (cocycle) if it contains an edge a such that $A - \{a\}$ is a cocycle (cycle). We say that A and a are *associated* with each other. From the above expression for $\{e\}$, we see that if e belongs to no bicycle, then $\{e\}$ is the sum of either a principal cycle $C(e)$ and a cocycle $D(e)$ or a cycle $C(e)$ and a principal cocycle $D(e)$. The former is the case if $e \in C(e)$ and the latter otherwise. The ordered pair $(C(e), D(e))$ is called a *decomposition* of e. Clearly, if B is any bicycle, then $C(e) + B$ is a cycle and $D(e) + B$ a cocycle; hence $(C(e) + B, D(e) + B)$ is another decomposition of e. Conversely, suppose that $(C_1(e), D_1(e))$ is another decomposition of e, and let $B = C(e) + C_1(e)$. Then B is a cycle. Since $\{e\} = C(e) + D(e) = C_1(e) + D_1(e)$, it follows that $D(e) + D_1(e) = C(e) + C_1(e) = B$, so that B is a bicycle. Thus, any decomposition of e is of the form $(C(e) + B, D(e) + B)$ for some bicycle B. Since $e \notin B$, we have $e \in C(e)$ if and only if $e \in C(e) + B$ and, similarly, $e \in D(e)$ if and only if $e \in D(e) + B$. Thus, we have the following generalisation of Corollary 8.14.

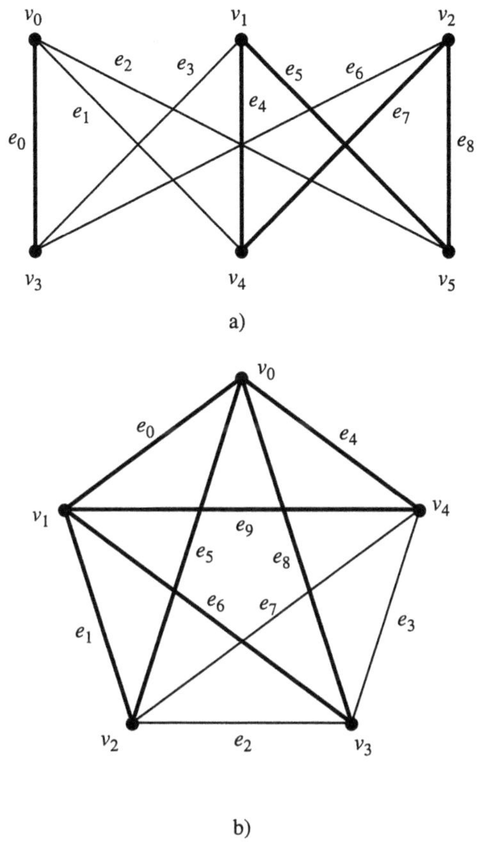

FIGURE 8.7.

Theorem 8.25 *For any edge e in a graph G, exactly one of the following possibilities holds: e is associated with a principal cycle or cocycle, or it belongs to a bicycle.*

Let P be the set of edges associated with principal cycles, let Q be the set of edges associated with principal cocycles, and let R be the union of the bicycles. Then (P, Q, R) is the *principal edge tripartition* of G. The edges of G are classified as *cyclic*, *cocyclic*, or *bicyclic* according to whether they belong to P, Q, or R. Note that every loop is cyclic and every isthmus cocyclic.

Example 8.6 *Consider $K_{3,3}$ and K_5 with the vertices and edges labelled as in Figure 8.7. In Figure 8.7a, the cocycle $\partial\{v_0, v_4, v_5\}$ is the set $D = \{e_0, e_4, e_5, e_7, e_8\}$, and $D - \{e_0\}$ is a cycle. Thus D is a principal cocycle, and so $e_0 \in Q$. By symmetry, the principal edge tripartition of $K_{3,3}$ is therefore $(\emptyset, EK_{3,3}, \emptyset)$. Likewise, in Figure 8.7b, the set $C = \{e_0, e_1, e_4, e_5, e_6, e_8, e_9\}$ is a cycle. Moreover, the cocycle $\partial\{v_0, v_1\}$ is $C - \{e_0\}$; hence C is a principal cycle. Therefore, $e_0 \in P$. By symmetry,* *it follows that the principal edge tripartition of K_5 is $(EK_5, \emptyset, \emptyset)$. Observe that neither $K_{3,3}$ nor K_5 has a non-empty bicycle.*

For each $e \in P$, we may write $\{e\} = C(e)+D(e)$, where $C(e)$ is a principal cycle and $D(e)$ is a cocycle. Thus $|C(e)\cap D(e)|$ is even. Since $|C(e)\cap D(e)| = |C(e)|-1$, the cardinality of any principal cycle is odd. Similarly, the cardinality of any principal cocycle is odd. A further conclusion is that the cardinality of any bicycle is even. In other words, $EG \in (\mathcal{C}(G) \cap \mathcal{C}^{\perp}(G))^{\perp}$. From Theorem 8.24, we may therefore write $EG = C + D$, where C is a cycle and D a cocycle. Thus, for any graph G there is a pseudo-partition of EG, of cardinality 2, whose elements are a cycle and a cocycle.

8.8 Diagonals

In preparation for our next characterisation of planar graphs, we must now generalise the concept of an algebraic diagonal.

A family $\mathcal{S}$ of closed walks in a graph G is said to be *balanced* if each edge of G is equal to just two components of walks in $\mathcal{S}$. If $\mathcal{S}$ is balanced, we sometimes refer to it as a *(walk) double cover*. An edge is *shared* with respect to $\mathcal{S}$ if it is a component of two distinct walks in $\mathcal{S}$.

Now let e be an edge that is not shared with respect to $\mathcal{S}$. Then there is a closed walk W in $\mathcal{S}$ that uses e just twice. We say that e is *$\mathcal{S}$-coherent* if e is a link and the boundary segments of W determined by e are cycles. (See Figure 8.8a.) Any other edge that is not shared with respect to $\mathcal{S}$ is said to be *$\mathcal{S}$-consistent*. (See Figure 8.8b.) Thus, any loop that is not shared is $\mathcal{S}$-consistent.

Next, for each component W of $\mathcal{S}$ let σ_W be a sense function for W. Let e be any edge of G. If e is shared with respect to $\mathcal{S}$, then e is a component of distinct walks X and Y in $\mathcal{S}$. Let $X = x_1x_2\ldots x_m$ and $Y = y_1y_2\cdots y_n$, and suppose that $e = x_i = y_j$. If $\sigma_X(i) = \sigma_Y(j)$, then we say that e is *$\mathcal{S}$-consistent*; otherwise, e is *$\mathcal{S}$-coherent*. (See Figures 8.9a and 8.9b.) Thus, the $\mathcal{S}$-consistency and $\mathcal{S}$-coherence

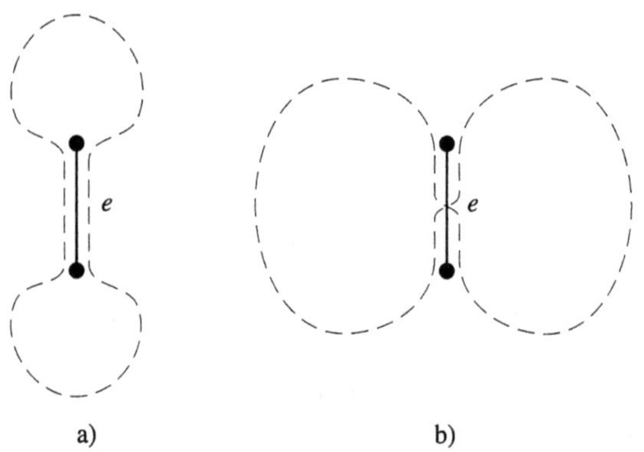

FIGURE 8.8.

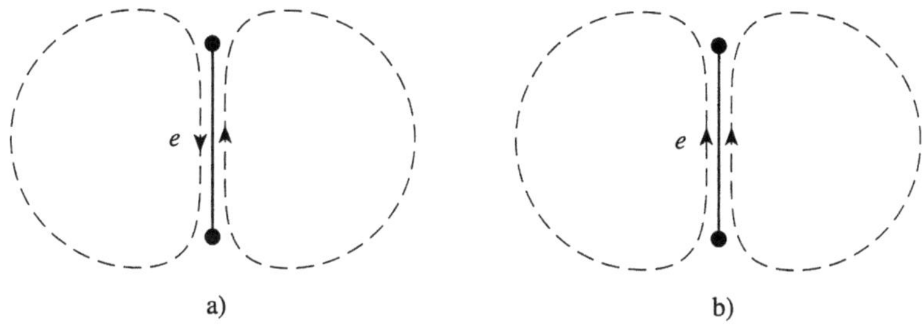

FIGURE 8.9.

of e depend not only on e and $\mathcal{S}$ but also on the sense functions chosen for the components of $\mathcal{S}$.

If I is an imbedding of G, then the edges that are skew-consistent and skew-coherent in I are those that are $\mathcal{S}$-consistent and $\mathcal{S}$-coherent, respectively, where $\mathcal{S}$ is the set of zigzag walks of I. In this way, we perceive $\mathcal{S}$-consistency and $\mathcal{S}$-coherence as generalisations of skew-consistency and skew-coherence of an imbedding.

Let $\mathcal{S}$ be a walk double cover of G. Let $e_0 W_0 e_1 W_1 \ldots e_{l-1} W_{l-1}$ be a non-empty string such that

(1) W_i is equivalent to a walk in $\mathcal{S}$ for each $i \in J_{l-1}^0$;

(2) $e_0, e_1, \ldots, e_{l-1}$ are distinct links of G;

(3) for each $i \in J_{l-1}^0$, e_i is a component of both W_{i-1} and W_i; and

(4) if $l = 1$, then $e_0 \notin \nabla W_0$.

(Here, and in what follows, subscripts are to be read modulo l.) Then $e_0 W_0 e_1 W_1 \cdots e_{l-1} W_{l-1}$ is a *ladder* $\mathcal{L}$ in $\mathcal{S}$ of *length* l. Each link e_i is a *rung* of $\mathcal{L}$. The ladder is *simple* if no two of its walks are equivalent. It is *straight* if the number of rungs that are $\mathcal{S}$-consistent is even; otherwise, the ladder is *twisted*. (Observe that this property is preserved under the replacement of W_i by W_i^* for any $i \in J_{l-1}^0$.) We denote the set of $\mathcal{S}$-coherent rungs of $\mathcal{L}$ by $\Omega(\mathcal{L})$.

For each $i \in J_{l-1}^0$, W_i is equivalent to a walk of the form $e_i X_i e_{i+1} X_i'$, where X_i and X_i' are walks. We then define $\sum_{i=0}^{l-1} \nabla X_i + \Omega(\mathcal{L})$ and $\sum_{i=0}^{l-1} \nabla X_i' + \Omega(\mathcal{L})$ as the *sides* of $\mathcal{L}$. These sides are described as *opposite* (each other).

Example 8.7 *Figure 8.10 (where vertices with the same label are identified) gives a twisted ladder $\mathcal{L}$ of length 6 in some walk double cover $\mathcal{S}$. The sense function for each walk may be inferred from the arrows. The bold lines give a side of $\mathcal{L}$. Note that $\mathcal{L}$ has three $\mathcal{S}$-consistent rungs and four $\mathcal{S}$-coherent ones.*

Let X be a side of the ladder $\mathcal{L} = e_0 W_0 e_1 W_1 \cdots e_{l-1} W_{l-1}$. Note that $X + \sum_{i=0}^{l-1} \nabla W_i$ is the side of $\mathcal{L}$ opposite X. For each $i \in J_{l-1}^0$, $X + \nabla W_i$ is a side of the ladder obtained by the replacement of W_i with W_i^*. Hence, the ladder $e_0 W_0^* e_1 W_1^* \cdots e_{l-1} W_{l-1}^*$ has the same sides as $\mathcal{L}$.

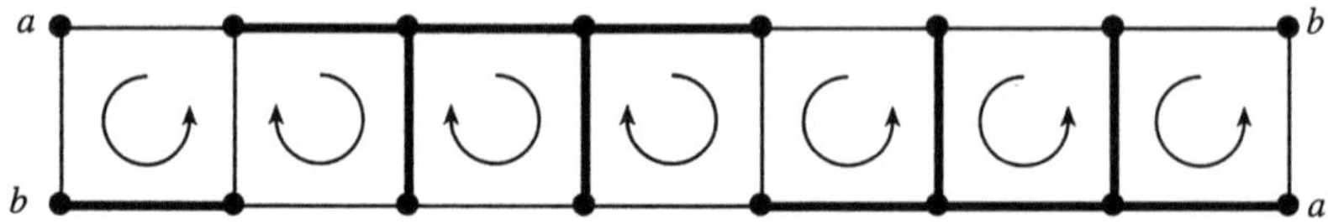

FIGURE 8.10.

A *diagonal* $\mathcal{D}$ in a graph G is a walk double cover such that the boundaries of the walks in $\mathcal{D}$ and the sides of the simple ladders in $\mathcal{D}$ are cocycles. Since the walks in $\mathcal{D}$ are closed, their boundaries must be cycles and hence bicycles. Thus every shared edge in $\mathcal{D}$ is bicyclic.

Let $\mathcal{L}$ be a ladder $e_0 W_0 e_1 W_1 \cdots e_{l-1} W_{l-1}$ in a walk double cover $\mathcal{S}$ of G, and suppose that ∇W_i is a cocycle of G for each $i \in J^0_{l-1}$. Let X be a side of $\mathcal{L}$ that is also a cocycle. Then the opposite side is a sum of cocycles, and hence again a cocycle. Moreover, if $\mathcal{L}'$ is obtained from $\mathcal{L}$ by the replacement of W_i with W_i^* for some $i \in J^0_{l-1}$, then the sides of $\mathcal{L}'$ are cocycles. In order to determine whether $\mathcal{S}$ is a diagonal, it therefore suffices to consider one side of a ladder of the form $e_0 W_0 e_1 W_1 \cdots e_{l-1} W_{l-1}$, where rungs $e_1, e_2, \ldots, e_{l-1}$ are $\mathcal{S}$-coherent.

Suppose that e is a loop in a graph G that has a diagonal $\mathcal{D}$. Since e is not a member of any cocycle of G, it is not contained in a boundary of a walk in $\mathcal{D}$ or in a side of a simple ladder in $\mathcal{D}$. Hence, some walk W in $\mathcal{D}$ uses e twice. Let W' be the walk obtained from W by deleting both occurrences of e, and let $\mathcal{D}' = (\mathcal{D} - \{W\}) \cup \{W'\}$. Then $\mathcal{D}'$ is clearly a diagonal of $G - \{e\}$.

Suppose that $\mathcal{L}$ is a ladder eW of length 1 in a diagonal $\mathcal{D}$ of G. Then e is a link that is used twice by the walk W. We may therefore write $W = eXeX'$, where X and X' are walks. Moreover, if e is $\mathcal{D}$-coherent, it follows that ∇X must be a cycle. (See Figure 8.8a.) Since $\nabla X + \{e\} = \nabla X \cup \{e\}$ is a side of $\mathcal{L}$, we also infer that in this case $\nabla X \cup \{e\}$ is a cocycle, in fact a principal cocycle, associated with e. Thus e is cocyclic. Similarly, if e is $\mathcal{D}$-consistent, then $\nabla X \cup \{e\}$ is a cycle and ∇X a cocycle so that e is cyclic. In summary, the unique rung in a ladder of length 1 is cocyclic if it is $\mathcal{D}$-coherent, and cyclic otherwise. Since any link that is not shared with respect to $\mathcal{D}$ is the unique rung of a ladder of length 1, each such edge is cyclic or cocyclic. Thus, an edge is bicyclic if and only if it is shared with respect to $\mathcal{D}$.

8.9 Every planar graph has a diagonal

We shall prove that a graph is planar if and only if it has a diagonal. This theorem generalises Rosenstiehl and Read's result that a connected graph with an algebraic diagonal and no non-empty bicycles is planar. To see this, suppose that W is an algebraic diagonal in a connected graph G with principal edge tripartition $(P, Q, \emptyset)$. Since W is a double tracing, ∇W is the empty cocycle. Moreover, any simple ladder in the walk double cover $\{W\}$ is of length 1. Let eW be such a simple ladder, and let the boundary segments of W determined by e be $A(e)$ and $B(e)$, respectively. If $e \in P$, then $A(e) \cup \{e\}$ is a cycle since W is an algebraic diagonal, and it follows that e is $\{W\}$-consistent. In this case, $A(e)$ is a side of eW and is a cocycle. On the other hand, if $e \in Q$, then $A(e)$ is a cycle and so e is $\{W\}$-coherent. Thus $A(e) \cup \{e\}$ is a side of eW and a cocycle. We conclude that $\{W\}$ is a diagonal and consequently that G is planar. The treatment follows Archdeacon et al. (1995).

Theorem 8.26 *Every planar graph has a diagonal.*

Proof. Let $\mathcal{D}$ be the family of zigzag walks of a planar imbedding of a graph G. We shall show that $\mathcal{D}$ is a diagonal of G. It is a walk double cover, and the boundary of each of its components is a cocycle by Lemma 8.16.

Let $\mathcal{L}$ be a simple ladder $e_0 W_0 e_1 W_1 \cdots e_{l-1} W_{l-1}$ in $\mathcal{D}$. As was noted before, we may suppose that rungs $e_1, e_2, \ldots, e_{l-1}$ are $\mathcal{D}$-coherent. Then e_0 is $\mathcal{D}$-coherent if and only if $\mathcal{L}$ is straight. For each $i \in J^0_{l-1}$, we write $W_i = e_i X_i e_{i+1} X'_i$. Let $X = \sum_{i=0}^{l-1} \nabla X_i + \{e_1, e_2, \ldots, e_{l-1}\}$. Then X is a side of $\mathcal{L}$ if $\mathcal{L}$ is twisted, and $X + \{e_0\}$ is a side of $\mathcal{L}$ otherwise. Let Y be the side of $\mathcal{L}$ that includes X. If $l = 1$, then the sides of $\mathcal{L}$ are cocycles by Theorem 8.22(1) or (2). Assume therefore that $l > 1$ and that any side of a ladder of length less than l is a cocycle.

Let $G' = G - \{e_1\}$, and let $\mathcal{D}' = (\mathcal{D} - \{W_0, W_1\}) \cup \{W'\}$, where $W' = e_0 X_0 X_1 e_2 X'_1 X'_0$. Clearly, $\mathcal{D}'$ is the family of zigzag walks in a planar imbedding of G'. Suppose that W' is equivalent to W_2. Then $l = 3$, since W_0 and W_{l-1} are the only components of $\mathcal{L}$ using e_0. Observe also that $\mathcal{L}$ is straight. (See Figure 8.11.) Moreover, the fact that W' and W_2 constitute the family of zigzag walks of an imbedding of G' forces EG' to be a circuit. It follows that $\{e_0, e_1, e_2\}$ is a cocycle of G. This cocycle is a side of $\mathcal{L}$. The argument is similar if W' is equivalent to W_{l-1}.

If W' is not equivalent to either W_2 or W_{l-1}, then

$$e_0 W' e_2 W_2 e_3 W_3 \cdots e_{l-1} W_{l-1}$$

is a simple ladder $\mathcal{L}'$ in G' of length $l - 1$. The edge e_0 is $\mathcal{D}'$-consistent if and only if it is $\mathcal{D}$-consistent and the edges $e_2, e_3, \ldots, e_{l-1}$ are $\mathcal{D}'$-coherent. Hence $Y - \{e_1\}$

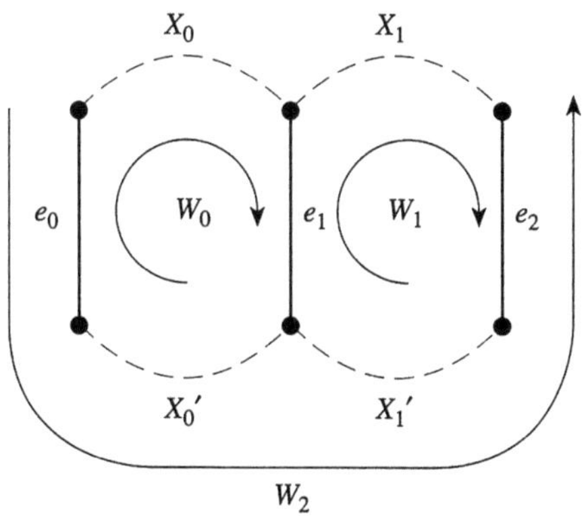

FIGURE 8.11.

is a side Y' of $\mathcal{L}'$. By hypothesis, Y' is a cocycle of G', and so Y or Y' is a cocycle of G.

If Y is a cocycle of G, we are done, and so we may assume that $Y - \{e_1\}$ is a cocycle of G. In fact, we may assume that $Y - \{e_i\}$ is a cocycle of G for each $i \in J^0_{l-1}$. Hence, for each i and j we find that $(Y - \{e_i\}) + (Y - \{e_j\}) = \{e_i, e_j\}$ is a cocycle of G. Any cycle of G containing rung e_i must therefore also contain e_j. But if $l > 2$, then ∇W_0 is a cycle of G containing e_0 and not e_2. We conclude that $l = 2$.

Thus $\mathcal{L}'$ is a ladder of length 1. If $\mathcal{L}$ is straight, then $e_0 \in Y'$ and $Y' - \{e_0\}$, being the boundary of a closed walk, is a cycle. Similarly, if $\mathcal{L}$ is twisted, then $e_0 \notin Y'$ and $Y' \cup \{e_0\}$ is a cycle. Since Y' is a cocycle of G, it follows that e_0 is cocyclic in the former case and cyclic in the latter. Both results contradict Theorem 8.25, for e_0 is bicyclic: It is in the boundaries of the distinct zigzag walks W_0 and W_1.

We conclude that Y is a cocycle of G and that $\mathcal{D}$ is a diagonal of G. ■

8.10 No non-planar graph has a diagonal

Our next goal is to prove the converse of Theorem 8.26 – that a non-planar graph cannot have a diagonal. The strategy is to use Wagner's theorem (Theorem 7.14, p.107). Specifically, we shall show that if a non-planar graph G has a diagonal, and G' is a graph obtained from G by the deletion or contraction of an edge, then G' has a diagonal. A contradiction is then established by observing that neither $K_{3,3}$ nor K_5 has a diagonal. We perform this task first.

The set of edges used by a walk W is called the *foundation* of W.

Lemma 8.27 *$K_{3,3}$ has no diagonal.*

Proof. We label the vertices and edges of $K_{3,3}$ as in Figure 8.7a. Suppose that there is a diagonal $\mathcal{D}$ in $K_{3,3}$. We show first that $\mathcal{D}$ has only one walk. Suppose not. Since $K_{3,3}$ has no non-empty bicycles, the boundary of each walk in $\mathcal{D}$ is empty. Thus the foundations of the walks in $\mathcal{D}$ form a partition of $EK_{3,3}$. There must be a vertex, say v_0, incident on edges in distinct cells of that partition. Without loss of generality, we may therefore suppose that some walk W in $\mathcal{D}$ uses e_0 but neither e_1 nor e_2. Thus $\{e_0\}$ is a side of the ladder e_0W in $\mathcal{D}$, since e_0 is $\mathcal{D}$-coherent. This result contradicts the fact that e_0 is not an isthmus.

Let D be the unique walk in $\mathcal{D}$. It uses every edge twice. Since D may be replaced by an equivalent walk if necessary, we may therefore write D as e_0Xe_0X', where e_0 and X are chosen to minimise the length of the walk X. X must use at least one edge of $\partial\{v_0, v_3\} = \{e_1, e_2, e_3, e_6\}$. If e_0 were $\mathcal{D}$-consistent, then $\nabla X \cup \{e_0\}$ would be a cycle. Moreover ∇X would be a side of the ladder e_0D and hence a cocycle. Thus e_0 would be cyclic, in contradiction to the fact that it is cocyclic. (See Example 8.6.) Hence e_0 must be $\mathcal{D}$-coherent. It follows that ∇X is a cycle and that $\nabla X \cup \{e_0\}$ is a cocycle, being a side of the ladder e_0D. This cocycle is therefore $\{e_0, e_4, e_5, e_7, e_8\}$, the principal cocycle associated with e_0, and hence contains none of e_1, e_2, e_3, e_6. We conclude that X uses one of these edges twice, in contradiction to the choice of e_0 and X. ■

Lemma 8.28 *K_5 has no diagonal.*

Proof. We label the vertices and edges of K_5 as in Figure 8.7b.

Let $\mathcal{D}$ be a diagonal in K_5. We show first that $\mathcal{D}$ has only one walk. Suppose not. Arguing as in the proof of Lemma 8.27, we find that the foundations of the walks in $\mathcal{D}$ partition EK_5. Without loss of generality, let v_0 be incident on edges in distinct cells of that partition. Each cell containing edges incident on a vertex v_i, where $i \in J_4^0$, has more than one such edge, for otherwise some edge incident on v_i would be an isthmus by the argument in the proof of Lemma 8.27. We may therefore suppose that $\mathcal{D}$ contains a walk D_1 using both e_0 and e_4 twice and a walk D_2 using both e_5 and e_8 twice. Thus e_0D_1 is a ladder $\mathcal{L}$ in $\mathcal{D}$. If e_0 is $\mathcal{D}$-consistent, then there is a side of $\mathcal{L}$ whose intersection with ∂v_0 is $\{e_4\}$. On the other hand, if e_0 is $\mathcal{D}$-coherent, then there is a side of $\mathcal{L}$ whose intersection with ∂v_0 is $\{e_0\}$. These sides must be cocycles, since $\mathcal{D}$ is a diagonal. Without loss of generality, we may therefore assume that ∂v_1 is a side of $\mathcal{L}$. Thus D_1 uses each edge of ∂v_1 as well as e_4. Since D_1 uses $e_1 \in \partial v_1$ and e_1 is not an isthmus, D_1 must also use a second edge incident on v_2. We now know that D_1 uses at least six edges; so must D_2 – a contradiction since $|EK_5| = 10 < 2 \cdot 6$.

Let D be the unique walk in $\mathcal{D}$. Without loss of generality, we write D as e_0Xe_0X'. If e_0 were $\mathcal{D}$-coherent, then ∇X would be a cycle and $\nabla X \cup \{e_0\}$ (a side of the ladder e_0D) would be a cocycle. This result contradicts the fact that e_0 is cyclic. (See Example 8.6.) Hence e_0 is $\mathcal{D}$-consistent, so that $\nabla X \cup \{e_0\}$ is a cycle and ∇X a cocycle. Thus $\nabla X \cup \{e_0\}$ must be $\{e_0, e_1, e_4, e_5, e_6, e_8, e_9\}$, the principal cycle associated with e_0. We may now write D as $e_0SabTbUe_0VaW$, where $a \in \partial\{v_0, v_1\}$, $b \in \{e_2, e_3, e_7\}$ and S, T, U, V, W are walks. But a, being adjacent to b, is in the principal cycle associated with b, and therefore must also appear in T. Hence $\mathcal{D}$ cannot exist. ■

We must now show that the property of having a diagonal is preserved by the operation of deletion or contraction of an edge e. These operations will be studied by means of three lemmas that cover the various ways in which e may be used by the walks in a diagonal. If e is a loop in a graph G with a diagonal, we have already shown that $G - \{e\}$ has a diagonal. Hence, we may assume that e is a link.

Lemma 8.29 *Let e be a link in a graph G with diagonal $\mathcal{D}$. Suppose that e is not shared in $\mathcal{D}$. Then $G - \{e\}$ has a diagonal if e is $\mathcal{D}$-consistent, and G_e has a diagonal if e is $\mathcal{D}$-coherent.*

Proof. The edge e is used twice by some walk D in $\mathcal{D}$. Hence, we may write $D = eX_1eX_2$, where X_1 and X_2 are walks. Define

$$\mathcal{D}' = (\mathcal{D} - \{D\}) \cup \{D'\},$$

where $D' = X_1 X_2^{-1}$. Then D' is a closed walk in the graph G', where $G' = G - \{e\}$ if e is $\mathcal{D}$-consistent and $G' = G_e$ if e is $\mathcal{D}$-coherent. Moreover, $\mathcal{D}'$ is a walk double cover of G'. We shall show that $\mathcal{D}'$ is a diagonal of G'.

First, observe that $\nabla D' = \nabla X_1 + \nabla X_2 = \nabla D$. Thus, the boundary of each walk in $\mathcal{D}'$ is a cocycle of G not containing e and hence a cocycle of G'.

Next we consider a simple ladder $\mathcal{L}'$ in $\mathcal{D}'$. With $\mathcal{L}'$ we associate a simple ladder $\mathcal{L}$ in $\mathcal{D}$ obtained by replacing D', if it appears in $\mathcal{L}'$, by D. By hypothesis, the sides of $\mathcal{L}$ are cocycles in G. If D' is not in $\mathcal{L}'$, then any side of $\mathcal{L}$ is also a side of $\mathcal{L}'$. Being a cocycle in G that does not contain e, it is also a cocycle in G'. lights-on-satilite.au sun-ra Suppose therefore that $\mathcal{L}'$ is of the form $e_0 W_0 e_1 W_1 \ldots e_{l-1} W_{l-1}$, and without loss of generality let $D' = W_0$. If e_0 and e_1 are both in X_1 or both in X_2, there is a side of $\mathcal{L}$ that does not contain e. Moreover, this side is identical to a side of $\mathcal{L}'$. It is a cocycle in G not containing e and hence a cocycle of G', as desired.

Without loss of generality, we may now suppose that $e_0 \in X_1$ and $e_1 \in X_2$. We observe that $\Omega(\mathcal{L}) + \Omega(\mathcal{L}') = \{e_1\}$. Moreover, if we write $D = eY_1 e_0 Y_2 e Y_3 e_1 Y_4$, where Y_1, Y_2, Y_3, Y_4 are walks, then $D' = Y_1 e_0 Y_2 Y_4^{-1} e_1 Y_3^{-1}$. (See Figure 8.12, where Figure 8.12a and Figure 8.12b illustrate the possibilities that e is $\mathcal{D}$-consistent or $\mathcal{D}$-coherent, respectively.) Putting $W_i = e_i Z_i e_{i+1} Z_i'$ for each $i \in J_{l-1}^0$, where Z_i and Z_i' are walks, we may assume without loss of generality that $\nabla(Y_2 e Y_3) + \sum_{i=1}^{l-1} \nabla Z_i + \Omega(\mathcal{L})$ and $\nabla(Y_2 Y_4) + \sum_{i=1}^{l-1} \nabla Z_i + \Omega(\mathcal{L}')$ are sides S and S' of $\mathcal{L}$ and $\mathcal{L}'$, respectively. Thus $e \notin S'$. Define

$$\begin{aligned} Y &= S + S' \\ &= \nabla(Y_2 e Y_3) + \nabla(Y_2 Y_4) + \{e_1\} \\ &= \nabla(Y_3 e_1 Y_4 e). \end{aligned}$$

Thus, if e is $\mathcal{D}$-coherent, then Y is a principal cocycle in G associated with e, since eD is a straight ladder of length 1 in G. Similarly, if e is $\mathcal{D}$-consistent, then $Y + \{e\}$ is a cocycle. By hypothesis, S is also a cocycle of G. It follows that either S' or $S' \cup \{e\}$ is a cocycle of G, the former being the case if e is $\mathcal{D}$-coherent. In both cases, S' is a cocycle of G', as required. ■

Lemma 8.30 *Let e be a link in a graph G with diagonal $\mathcal{D}$. Suppose that e is not shared in $\mathcal{D}$. Then G_e has a diagonal if e is $\mathcal{D}$-consistent and $G - \{e\}$ has a diagonal if e is $\mathcal{D}$-coherent.*

Proof. The edge e is used twice by some walk D in a diagonal $\mathcal{D}$ of G. Hence we may write $D = eX_1 e X_2$, where X_1 and X_2 are walks. Let $G' = G_e$ if e is $\mathcal{D}$-consistent and $G' = G - \{e\}$ if e is $\mathcal{D}$-coherent. Observe that X_1 and X_2 are closed walks in G'. Define

$$\mathcal{D}' = (\mathcal{D} - \{D\}) \cup \{X_1, X_2\}.$$

Then $\mathcal{D}'$ is a double cover of G'. We shall show that $\mathcal{D}'$ is a diagonal of G'.

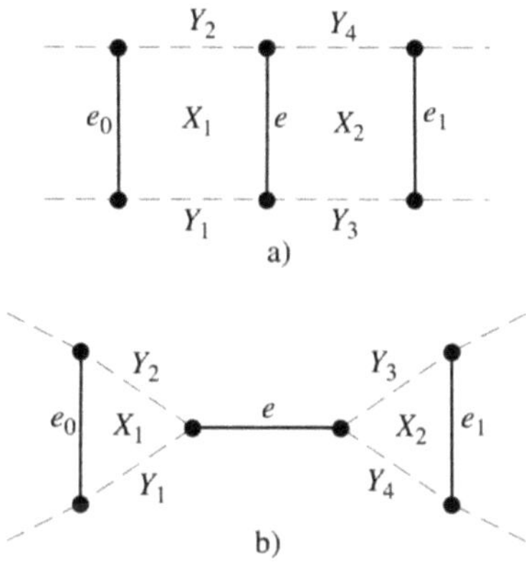

FIGURE 8.12.

First we show that the boundary of each walk D' in $\mathcal{D}'$ is a cocycle of G'. The boundary of each walk in $\mathcal{D}$ is a cocycle in G that does not contain e,and hence is a cocycle in G'. It remains to show that ∇X_1 and ∇X_2 are cocycles in G'. If e is $\mathcal{D}$-consistent, then eD is a twisted ladder in the diagonal $\mathcal{D}$ of G. It follows that ∇X_1 and ∇X_2 are cocycles of G. Since they do not contain e, they are also cocycles of $G_e = G'$. Similarly, if e is $\mathcal{D}$-coherent, then eD is a straight ladder, and $\nabla(X_1 e)$ and $\nabla(X_2 e)$ are cocycles of G. Hence ∇X_1 and ∇X_2 are cocycles of $G - \{e\} = G'$.

Next we need to show that any side of a simple ladder $\mathcal{L}'$ in $\mathcal{D}'$ is a cocycle in G'. We associate a ladder $\mathcal{L}$ in $\mathcal{D}$ with $\mathcal{L}'$ by replacing each of X_1 and X_2, if they appear in $\mathcal{L}'$, by D.

Note first that if neither X_1 nor X_2 is in $\mathcal{L}'$, then any side of $\mathcal{L}'$ is a cocycle in G that does not contain e and hence a cocycle in G'.

Next suppose that exactly one of X_1 and X_2 is in $\mathcal{L}'$. Then $\mathcal{L}$ is simple. Moreover, one of the sides of $\mathcal{L}$ is identical to a side of $\mathcal{L}'$. It is a cocycle of G not containing e and hence a cocycle of G'.

However, it may be that both X_1 and X_2 are in $\mathcal{L}'$. Suppose therefore that $\mathcal{L}' = e_0 W_0 e_1 W_1 \cdots e_{l-1} W_{l-1}$, and without loss of generality let $X_1 = W_0$. Then $X_2 = W_j$ for some $j \in J_{l-1}$. Let $W_i = e_i Z_i e_{i+1} Z_i'$ for each $i \in J_{l-1}^0$, where Z_i and Z_i' are walks. We distinguish two cases.

Case 1. Suppose that $1 < j < l - 1$. Then $e_1 W_1 e_2 W_2 \cdots e_{j-1} W_{j-1} e_j D$ and $e_{j+1} W_{j+1} e_{j+2} W_{j+2} \ldots e_{l-1} W_{l-1} e_0 D$ are simple ladders, $\mathcal{L}_1$ and $\mathcal{L}_2$, respectively, in $\mathcal{D}$. Thus $\Omega(\mathcal{L}_1) + \Omega(\mathcal{L}_2) = \Omega(\mathcal{L}')$. In order to find sides of $\mathcal{L}_1$, $\mathcal{L}_2$, and $\mathcal{L}'$, we observe first that D may be written without loss of generality as $Y_1 e_0 Y_2 e_1 Y_3 e Y_4 e_{j+1} Y_5 e_j Y_6 e$ or $Y_1 e_0 Y_2 e_1 Y_3 e Y_4 e_j Y_5 e_{j+1} Y_6 e$, where Y_1, Y_2, Y_3, Y_4, Y_5, Y_6 are walks.

Subcase 1.1. Suppose that $D = Y_1 e_0 Y_2 e_1 Y_3 e Y_4 e_{j+1} Y_5 e_j Y_6 e$. (See Figure 8.13, where Figure 8.13a and Figure 8.13b illustrate the possibilities that e is $\mathcal{D}$-consistent

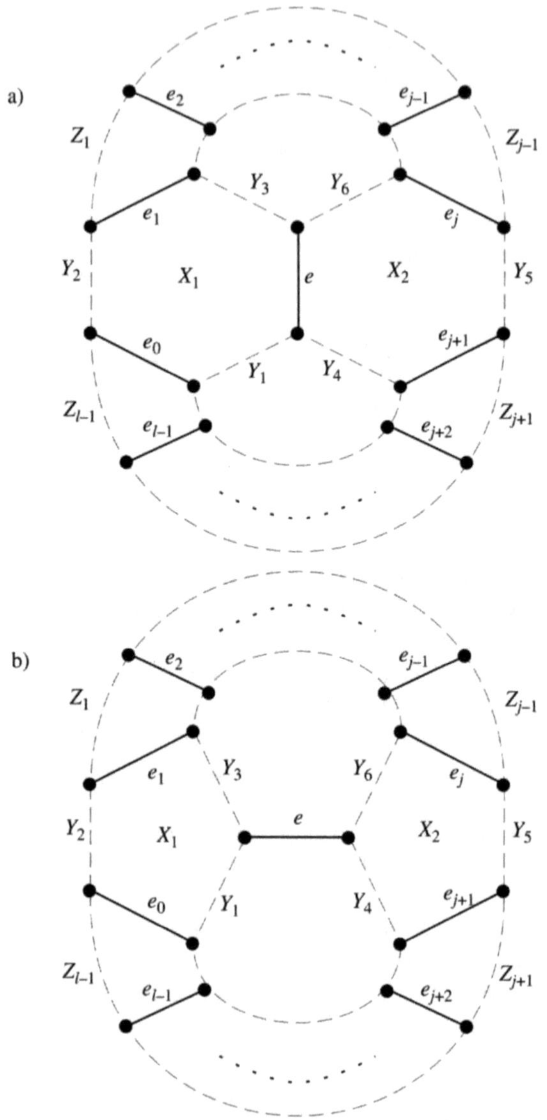

FIGURE 8.13.

or $\mathcal{D}$-coherent, respectively.) Then

$$\nabla Y_2 + \sum_{i=1}^{j-1} \nabla Z_i + \nabla(Y_6 Y_4) + \sum_{i=j+1}^{l-1} \nabla Z_i + \Omega(\mathcal{L}')$$

is a side S' of $\mathcal{L}'$,

$$\nabla(Y_6 e Y_1 e_0 Y_2) + \sum_{i=1}^{j-1} \nabla Z_i + \Omega(\mathcal{L}_1)$$

is a side S_1 of $\mathcal{L}_1$, and

$$\nabla(Y_2e_1Y_3eY_4) + \sum_{i=j+1}^{l-1} \nabla Z_i + \Omega(\mathcal{L}_2)$$

is a side S_2 of $\mathcal{L}_2$. Define

$$\begin{aligned} Z &= S' + S_1 + S_2 + \nabla(Y_1e_0Y_2e_1Y_3) \\ &= \nabla Y_2 + \nabla(Y_6Y_4) + \nabla(Y_6eY_1e_0Y_2) + \\ &\quad \nabla(Y_2e_1Y_3eY_4) + \nabla(Y_1e_0Y_2e_1Y_3) \\ &= \emptyset. \end{aligned}$$

If e is $\mathcal{D}$-consistent, then $\nabla(Y_1e_0Y_2e_1Y_3)$ is a cocycle in G, as are S_1 and S_2. Hence S' is a cocycle of G and therefore of G'. Suppose now that e is $\mathcal{D}$-coherent. This time, $\nabla(Y_1e_0Y_2e_1Y_3e)$ is a cocycle of G. It follows that $S' + \{e\}$ is a cocycle of G so that again S' is a cocycle of G'.

Subcase 1.2. Suppose that $D = Y_1e_0Y_2e_1Y_3eY_4e_jY_5e_{j+1}Y_6e$. This time,

$$\nabla Y_2 + \sum_{i=1}^{j-1} \nabla Z_i + \nabla Y_5 + \sum_{i=j+1}^{l-1} \nabla Z_i + \Omega(\mathcal{L}')$$

is a side S' of $\mathcal{L}'$,

$$\nabla(Y_5e_{j+1}Y_6eY_1e_0Y_2) + \sum_{i=1}^{j-1} \nabla Z_i + \Omega(\mathcal{L}_1)$$

is a side S_1 of $\mathcal{L}_1$, and

$$\nabla(Y_2e_1Y_3eY_4e_jY_5) + \sum_{i=j+1}^{l-1} \nabla Z_i + \Omega(\mathcal{L}_2)$$

is a side S_2 of $\mathcal{L}_2$. Define

$$\begin{aligned} Z &= S' + S_1 + S_2 \\ &= \nabla Y_2 + \nabla Y_5 + \nabla(Y_5e_{j+1}Y_6eY_1e_0Y_2) + \nabla(Y_2e_1Y_3eY_4e_jY_5) \\ &= \nabla D. \end{aligned}$$

Then Z is a cocycle of G not containing e and hence a cocycle of G'. Moreover, S_1 and S_2 are cocycles of G containing e, so that $S_1 + S_2$ is a cocycle of G missing e and hence a cocycle of G'. It follows that S' is a cocycle of G', as desired.

Case 2. Without loss of generality, we may now suppose that $j = 1$. Note that $e_0De_2W_2e_3W_3 \cdots e_{l-1}W_{l-1}$ is a single ladder $\mathcal{M}$ in $\mathcal{D}$. Without loss of generality, we may take D to be $Y_1e_0Y_2e_1Y_3eY_4e_1Y_5e_2Y_6e$ or $Y_1e_0Y_2e_1Y_3eY_4e_2Y_5e_1Y_6e$, where $Y_1, Y_2, Y_3, Y_4, Y_5, Y_6$ are walks.

Subcase 2.1. Suppose that $D = Y_1e_0Y_2e_1Y_3eY_4e_1Y_5e_2Y_6e$. Then

$$\nabla Y_2 + \nabla Y_5 + \sum_{i=2}^{l-1} \nabla Z_i + \Omega(\mathcal{L}')$$

is a side S' of $\mathcal{L}'$ and

$$\nabla(Y_2e_1Y_3eY_4e_1Y_5) + \sum_{i=2}^{l-1} \nabla Z_i + \Omega(\mathcal{M})$$

is a side T of $\mathcal{M}$. Put $Z = S' + T$. If e_1 is $\mathcal{D}$-consistent, then

$$\begin{aligned} Z &= \nabla Y_2 + \nabla Y_5 + \nabla(Y_2e_1Y_3eY_4e_1Y_5) \\ &= \nabla(Y_3eY_4). \end{aligned}$$

If e_1 is $\mathcal{D}$-coherent, then $e_1 \in S' - T$, and so $Z = \nabla(Y_3eY_4e_1)$. But in any case Z is a cocycle, since e_1D is a ladder in $\mathcal{D}$. Moreover, T is a cocycle of G by assumption. Hence S' is a cocycle of G not containing e and therefore a cocycle of G', as required.

Subcase 2.2. Suppose that $D = Y_1e_0Y_2e_1Y_3eY_4e_2Y_5e_1Y_6e$. First observe that

$$\nabla Y_2 + \nabla(Y_6Y_4) + \sum_{i=2}^{l-1} \nabla Z_i + \Omega(\mathcal{L}')$$

is a side S' of $\mathcal{L}'$ and that

$$\nabla(Y_2e_1Y_3eY_4) + \sum_{i=2}^{l-1} \nabla Z_i + \Omega(\mathcal{M})$$

is a side T of $\mathcal{M}$. Put $Z = S' + T$. If e_1 is $\mathcal{D}$-consistent, then

$$\begin{aligned} Z &= \nabla Y_2 + \nabla(Y_6Y_4) + \nabla(Y_2e_1Y_3eY_4) \\ &= \nabla(e_1Y_3eY_6). \end{aligned}$$

If e_1 is $\mathcal{D}$-coherent, then $Z = \nabla(Y_3eY_6)$.

Since e_1D is a ladder of $\mathcal{D}$, it follows that $\nabla(Y_3eY_4e_2Y_5)$ is a cocycle if e_1 is $\mathcal{D}$-consistent and that $\nabla(Y_3eY_4e_2Y_5e_1)$ is a cocycle if e_1 is $\mathcal{D}$-coherent. In both cases, the sum Z' of this cocycle and Z is $\nabla(Y_4e_2Y_5e_1Y_6)$. Thus Z' or $Z' + \{e\}$ is a cocycle of G (depending on whether e is $\mathcal{D}$-consistent or $\mathcal{D}$-coherent), since eD is a ladder of $\mathcal{D}$. Hence Z is a cocycle of G if e is $\mathcal{D}$-consistent, and $Z + \{e\}$ is a cocycle of G if e is $\mathcal{D}$-coherent. Because the side T of $\mathcal{M}$ is a cocycle of G by assumption, it follows that S' or $S' + \{e\}$ is a cocycle of G, depending on whether e is $\mathcal{D}$-consistent or $\mathcal{D}$-coherent. Since $e \notin S'$, in both cases S' is a cocycle of G', as required. ■

Lemma 8.31 *Let G be a graph with a diagonal $\mathcal{D}$, and let e be a shared edge with respect to $\mathcal{D}$. Then $G - \{e\}$ and G_e have diagonals.*

Proof. Let D_1 and D_2 be the two walks containing e in a diagonal $\mathcal{D}$ of G. We may write $D_1 = eX_1$ and $D_2 = eX_2$. We may also suppose without loss of generality that e is $\mathcal{D}$-consistent, since D_1 may be replaced by D_1^* if necessary. Define

$$\mathcal{D}' = (\mathcal{D} - \{D_1, D_2\}) \cup \{D'\},$$

where $D' = X_1X_2^{-1}$ if e is to be deleted and $D' = X_1X_2$ if e is to be contracted. In either case, $\mathcal{D}'$ is a double cover of G', where $G' = G - \{e\}$ or $G' = G_e$. We shall show that $\mathcal{D}'$ is a diagonal of G'.

Note first that $\nabla D' = \nabla D_1 + \nabla D_2$. Thus, the boundary of each member of $\mathcal{D}'$ is a cocycle of G not containing e and hence a cocycle of G'.

Next, let $\mathcal{L}'$ be a simple ladder in $\mathcal{D}'$. If D' is not in $\mathcal{L}'$, then $\mathcal{L}'$ is also a ladder in G. Its sides are therefore cocycles in G not containing e and hence are cocycles in G'. We may therefore assume that D' is in $\mathcal{L}'$. Without loss of generality, let $\mathcal{L}' = e_0D'e_1W_1 \cdots e_{l-1}W_{l-1}$.

If both e_0 and e_1 are in X_1, let $\mathcal{L}$ be the ladder in G formed from $\mathcal{L}'$ by replacing D' with D_1. Then one side of $\mathcal{L}$ is identical to a side of $\mathcal{L}'$. Since it is a cocycle of G not containing e, it is a cocycle of G', as required. A similar argument obtains if e_0 and e_1 are in X_2.

Suppose therefore that e_0 is in X_1 and e_1 is in X_2. This time, we define $\mathcal{L}$ as the ladder of length $l + 1$ in $\mathcal{D}$ obtained from $\mathcal{L}'$ by replacing D' with D_1eD_2. Letting S' be a side of $\mathcal{L}'$, we find that S' is a side of $\mathcal{L}$ if e is to be contracted, since e is $\mathcal{D}$-consistent and $D' = X_1X_2$. In this case, S' is a cocycle of G not containing e and hence a cocycle of G'. If e is to be deleted, then $D' = X_1X_2^{-1}$, and $S' + \{e\}$ is a side of the ladder obtained from $\mathcal{L}$ by the replacement of D_2 with D_2^*. Hence $S' + \{e\}$ is a cocycle of G and S' a cocycle of G'. ∎

Thus, we now have the following theorem.

Theorem 8.32 *A graph is planar if and only if it has a diagonal.*

An alternative approach to the proof of this theorem has been given by Keir and Richter (to appear).

Lemma [illegible] *graph* [illegible] *diagonal P, and let* [illegible] *edge with respect to P. Then* $G - \{e\}$ *and* G_e [illegible] *diagonals.*

[illegible]

[illegible]

9

Rings of Bonds

In this chapter, we give another characterisation of planar graphs, due to Chernyak (1980). In fact, we shall prove a general theorem about consistently orientable bonds in an arbitrary graph. The characterisation of planar graphs is obtained by dualising this theorem. The argument depends on a property of chordal graphs. Accordingly, chordal graphs are defined and studied in the next section.

9.1 Chordal graphs

Recall that a circuit C in a graph G is induced if $G[VC] = G[C]$. A *chordal* graph is a simple graph with no induced circuits of length greater than 3.

Let S be a minimal set of vertices in a graph G with the property that $G - S$ has more components than G. Then S is a *cutset* of G.

We need the following theorem, due to Dirac (1961).

Theorem 9.1 *Let S be a cutset in a chordal graph G. Then any two distinct vertices of S are adjacent.*

Proof. Clearly, the vertices of S belong to the same component of G, and so we may assume that G is connected.

Suppose that there exist distinct non-adjacent vertices u and v in S. Let H_1 and H_2 be distinct components of $G - S$. Note that u is joined to a vertex $x_1 \in VH_1$ by an edge e_1, for otherwise $G - (S - \{u\})$ is not connected, in contradiction to the minimality of S. Similarly that v is joined to a vertex $y_1 \in VH_1$ by an edge a_1, and u and v are joined to vertices x_2 and y_2 in VH_2 by edges e_2 and

a_2, respectively. For each $i \in J_2$, let P_i be a path in H_i, of minimum length, joining x_i and y_i. We may assume that x_i and y_i are chosen to minimise $|P_i|$. The minimality of $|P_i|$ ensures that vertices of VP_i that are not adjacent in $G[P_i]$ are not adjacent in H_i either. Moreover, neither u nor v can be adjacent to any vertex of $(VP_1 \cup VP_2) - \{x_1, x_2, y_1, y_2\}$. It follows that $P_1 \cup P_2 \cup \{e_1, e_2, a_1, a_2\}$ is an induced circuit C of G. Since $|C| \geq 4$, we have a contradiction to the fact that G is chordal. ■

9.2 Rings of bonds

In this section, we prove that any strict maximal ring of bonds in a graph is even. First, we show how this result yields another characterisation of planar graphs. The mechanism is the concept of duality. Specifically, we apply the following theorem.

Theorem 9.2 *Let G^D be the dual of a planar graph G with respect to a planar imbedding of G. Suppose that there exists a strict maximal ring $\mathcal{S}$ of circuits in G. Then $\mathcal{S}^D$ is a strict maximal ring of bonds in G^D.*

Proof. Let $\mathcal{S} = \{C_0, C_1, \ldots, C_{n-1}\}$, and suppose that, for each $i \in J^0_{n-1}$, $C_{i-1} \cap C_i \cap C_{i+1} = \emptyset$ and C_i meets both C_{i-1} and C_{i+1} but no other circuit of $\mathcal{S}$. (Throughout this section, all subscripts are read modulo n.) Then $\mathcal{S}^D = \{C^D_0, C^D_1, \ldots, C^D_{n-1}\}$. We shall show that $\mathcal{S}^D$ is the required ring of bonds in G^D. Clearly $|\mathcal{S}^D| = |\mathcal{S}| \geq 3$, and each element of $\mathcal{S}^D$ is a bond of G^D. Note also that an edge belongs to circuits X and Y in G if and only if it dualises to an edge of $X^D \cap Y^D$ in G^D. It follows that $C^D_{i-1} \cap C^D_i \cap C^D_{i+1} = \emptyset$ for each i, and C^D_i meets both C^D_{i-1} and C^D_{i+1} but no other bond of $\mathcal{S}^D$. In order to show that $\mathcal{S}^D$ is a ring, it remains only to establish that its bonds are consistently orientable.

Since $\mathcal{S}$ is consistently orientable, we may assume as given an orientation ρ of G under which each circuit of $\mathcal{S}$ is directed. Let K be a gem corresponding to a planar imbedding of G. Thus K is bipartite. Recall that $\mathcal{R}(K)$ and $\mathcal{Y}(K)$ are the sets of blue-yellow and red-blue bigons, respectively, in K. By Theorem 4.6, $\mathcal{R}(K) \cup \mathcal{Y}(K)$ is consistently orientable. Let σ be a consistent orientation for $\mathcal{R}(K) \cup \mathcal{Y}(K)$. Then σ is an orientation of K. Now choose a link $e \in EG$ joining vertices u and v, where $\rho(e) = v$. Recall that u and e are bigons in K. Call these bigons M and N, respectively. Then N is a red-blue square. We obtain a new orientation of K from σ by changing the orientation of the unique (red) edge of $M \cap N$. Under this orientation of K, the red edges of N form a directed bond of the graph $K[N]$. By performing this operation for each link $e \in EG$ in turn, we obtain an orientation σ^D of K. This orientation is also an orientation of K^D. For each red-blue bigon Z of K^D that is not an isthmus of G^D, it follows that the blue edges of Z form a directed bond of $K^D[Z]$ under σ^D.

Let d be a blue edge of K^D, and let d join vertices x and y, where $\sigma^D(d) = y$. Let x and y be vertices of red-yellow bigons A and B, respectively, in K^D. These bigons are vertices of G^D. Moreover, d is contained in a red-blue bigon Y of K^D, and Y

is an edge of G^D joining A and B. If $Y \in E(\text{Fnd}(G^D))$, we define $\rho^D(Y) = B$. In this case, $\rho^D(Y)$ is well defined, since the blue edges of Y form a directed bond of $K^D[Y]$ under σ^D. The application of this definition to each member of $E(\text{Fnd}(G^D))$ yields an orientation of $\text{Fnd}(G^D)$, which may be extended to an orientation ρ^D of G^D by defining $\rho^D(b)$ to be either end of b for each isthmus b of G^D.

Next, we show that any circuit C in G that is directed under ρ dualises to a bond C^D in G^D that is directed under ρ^D. C corresponds to a semicycle X in K that contains, at most, one blue edge of each red-blue bigon in K. Since K is bipartite, we may choose X to be normal. In fact, by adding to X appropriate red-blue bigons if necessary, we can obtain a semicycle X' that represents C and has the property that no blue edge of X' is adjacent to a red edge of X'. We may therefore assume that X has this property.

Now consider a particular edge $e_1 \in C$. It corresponds to a red-blue bigon Y_1 in K that contains a unique (blue) edge of X. Since K is spherical, we may write $X = \sum \mathcal{U}$, where $\mathcal{U}$ is a set of bigons of K. Let $\mathcal{R}$ be the set of blue-yellow bigons in $\mathcal{U}$. Thus $\mathcal{R} \subseteq V_G^D$. Of the two blue-yellow bigons in K that meet Y_1, exactly one is contained in $\mathcal{R}$. (These bigons are, of course, red-yellow bigons in K^D.) Let a_1 be a red edge in Y_1. (It is therefore a blue edge in K^D.) Since $\sum(\mathcal{R}(K) \cup \mathcal{B}(K) \cup \mathcal{Y}(K)) = \emptyset$, without loss of generality we may assume that $\sigma^D(a_1)$ is a vertex of a red-yellow bigon in K^D that is not in $\mathcal{R}$. It follows that $\rho^D(e_1) \notin \mathcal{R}$.

Next, consider the edge $e_2 \in C - \{e_1\}$ incident on $\rho(e_1)$. It corresponds to a red-blue bigon Y_2 in K. Of the two blue-yellow bigons in K that meet Y_2, exactly one is contained in $\mathcal{R}$. Let a_2 be a blue edge of K^D in Y_2. Since C is a directed circuit under ρ, by construction $\sigma^D(a_2)$ is also a vertex of a red-yellow bigon in K^D that is not in $\mathcal{R}$. It follows that $\rho^D(e_2) \notin \mathcal{R}$.

Proceeding inductively, we infer that $\rho^D(c) \notin \mathcal{R}$ for each edge $c \in C$. Therefore C^D is a directed bond under ρ^D. Since the circuits of $\mathcal{S}$ are consistently orientable, it now follows that the bonds of $\mathcal{S}^D$ are consistently orientable. Hence $\mathcal{S}^D$ is a ring.

It is clear that if $\mathcal{S}^D$ were not maximal, then $\mathcal{S}$ would not be maximal either. We conclude that $\mathcal{S}^D$ is maximal.

It remains to show that $\mathcal{S}^D$ is strict. Suppose not. Then there are circuits $C_i \in \mathcal{S}$ and $C_j \in \mathcal{S}$ such that $C_i \cap C_j = \emptyset$ and $C_i \cup C_j$ includes a circuit distinct from C_i and C_j. The last condition implies that $|VC_i \cap VC_j| > 1$, in contradiction to the facts that $C_i \cap C_j = \emptyset$ and $\mathcal{S}$ is strict. Hence $\mathcal{S}^D$ is strict. ■

Thus, the main theorem of this section implies that any strict maximal ring of circuits in a planar graph must be even.

First, we show that any non-planar graph has a strict maximal odd ring of circuits. By Kuratowski's theorem, it suffices to establish this result for K_5 and $K_{3,3}$.

Lemma 9.3 *$K_{3,3}$ has a strict maximal 3-ring of circuits.*

Proof. Let the vertices and edges of $K_{3,3}$ be labelled as in Figure 8.7a (p. 129). Let $\mathcal{S}$ be the strict 3-ring

$$\{\{e_0, e_1, e_4, e_3\}, \{e_4, e_5, e_8, e_7\}, \{e_0, e_2, e_8, e_6\}\}.$$

We must show that $\mathcal{S}$ is maximal. In any ring $\mathcal{S}'$ of cardinality more than 3, the sum of the lengths of the circuits is at least $4 \cdot 4 = 16 > 12 = 6 \cdot 2$. Therefore, some vertex v of $K_{3,3}$ has at least three circuits of $\mathcal{S}'$ passing through it. If $K_{3,3}$ is oriented so that each circuit of $\mathcal{S}'$ is a directed circuit, then either the indegree or the outdegree of v is 1. Hence, at least three circuits of $\mathcal{S}'$ contain some edge e incident on v, in contradiction to the definition of a ring. ■

Lemma 9.4 *K_5 has a strict maximal 5-ring of circuits.*

Proof. Let the vertices and edges of K_5 be labelled as in Figure 8.7b (p. 129). Let $\mathcal{S}$ be the strict 5-ring

$$\{\{e_0, e_1, e_5\}, \{e_1, e_2, e_6\}, \{e_2, e_3, e_7\}, \{e_3, e_4, e_8\}, \{e_4, e_0, e_9\}\}.$$

We must show that $\mathcal{S}$ is maximal. Suppose that $\mathcal{S}'$ is a ring $\{C_0, C_1, \ldots, C_{k-1}\}$ such that $k \geq 6$, where $C_i \cap C_j \neq \emptyset$ if and only if $i = j$ or $i \equiv j \pm 1 (\mathrm{mod}\ k)$. The sum of the lengths of the circuits of $\mathcal{S}'$ is therefore at least $6 \cdot 3 = 18$.

Since any edge of K_5 belongs to, at most, two circuits of $\mathcal{S}'$, at most four circuits of $\mathcal{S}'$ pass through a given vertex. If there were fewer than three vertices with four circuits of $\mathcal{S}'$ passing through them, then the sum of the lengths of the circuits of $\mathcal{S}'$ would be, at most, $2 \cdot 4 + 3 \cdot 3 = 17$. Thus, we may assume that v_0, v_1, v_2 each have four circuits of $\mathcal{S}'$ passing through them. Every edge incident on these vertices belongs to two circuits of $\mathcal{S}'$. If K_5 is oriented so that every circuit of $\mathcal{S}'$ is a directed circuit, then exactly one or two of the three edges of $EK_5 - \{e_3\}$ incident on v_3 are directed to v_3, for otherwise e_3 would carry more than two circuits of $\mathcal{S}'$. Therefore, exactly four circuits of $\mathcal{S}'$ pass through v_3. By symmetry, the same result holds for v_4. Thus, every edge of K_5 belongs to two circuits of $\mathcal{S}'$, and the sum of the lengths of the circuits of $\mathcal{S}'$ is $5 \cdot 4 = 20$.

If three or more circuits of $\mathcal{S}'$ were of length 4 or greater, then the sum of the lengths of the circuits of $\mathcal{S}'$ would be at least $4 \cdot 3 + 3 \cdot 3 = 21$. Therefore, at most, two circuits of $\mathcal{S}'$ have length greater than 3. Since $|\mathcal{S}'| \geq 6$, it follows that $\mathcal{S}'$ contains two triangles that meet. Since K_5 has no digons, these triangles have just one edge in common. Let these triangles be $C_1 = \{e_0, e_1, e_5\}$ and $C_2 = \{e_1, e_2, e_6\}$. Since e_0 belongs to two circuits of $\mathcal{S}'$, there must be another edge e incident on v_1 such that $\{e_0, e\} \subset C_0$. Similarly, there must exist an edge $e' \neq e_6$ incident on v_1 such that $\{e', e_6\} \subset C_3$. Since the degree of v_1 is 4, we must have $e' = e$. Hence $C_0 \cap C_3 \neq \emptyset$, in contradiction to the fact that $\mathcal{S}'$ is a ring with $|\mathcal{S}'| \geq 6$. Therefore, $\mathcal{S}$ must be maximal, as required. ■

Let $\{B_0, B_1, \ldots, B_{n-1}\}$ be a ring of bonds, where $B_i \cap B_j \neq \emptyset$ if and only if $i = j$ or $i \equiv j \pm 1 (\mathrm{mod}\ n)$. For any $i \in J_{n-1}^0$, we have $B_i \cap B_{i-2} = \emptyset$ if and only if $n \geq 4$. When proving that no graph has a strict maximal odd ring of bonds, we therefore treat a 3-ring as a special case.

Lemma 9.5 *No graph has a maximal* 3*-ring of bonds.*

Proof. Suppose that there exists a graph G that has a maximal 3-ring $\{B_0, B_1, B_2\}$ of bonds. Choose G to minimise $|EG|$. Thus G is connected.

Should there exist $e \in EG - (B_0 \cup B_1 \cup B_2)$, then $\{B_0, B_1, B_2\}$ would remain a maximal 3-ring in G_e, in contradiction to the choice of G. We conclude that $B_0 \cup B_1 \cup B_2 = EG$.

Next, we show that G is chordal. Suppose that G has an induced circuit C of length $m \geq 4$. Let $VC = \{v_1, v_2, \dots, v_m\}$, where v_i is adjacent to v_{i+1} for each $i < m$. If m is even, then $\{\partial v_1, \partial v_2, \dots, \partial v_m\}$ is an m-ring. If m is odd, then $\{\partial v_1, \partial v_2, \dots, \partial v_{m-2}, \partial\{v_{m-1}, v_m\}\}$ is an $(m-1)$-ring. In either case, the maximality of $\{B_0, B_1, B_2\}$ is contradicted. Hence G is chordal.

Our next goal is to show that no subgraph of G is isomorphic to K_4. Let H be a subgraph of G isomorphic to K_4. We may assume that G is oriented so that B_0, B_1, B_2 are directed bonds. Without loss of generality, we may also assume that one of the following possibilities holds:

(1) H has neither a source nor a sink;

(2) H has a unique source but no sink;

(3) H has a unique source and a unique sink.

All of the non-isomorphic possibilities are illustrated in Figure 9.1. We adopt the labelling of the edges given in that figure. Note that if $B_i \cap EH \neq \emptyset$ for some i, then $B_i \cap EH$ is a square or a vertex cocycle of H. In cases (1) and (2), none of the three squares can be a subset of a directed bond. Therefore, the edge b, for example, belongs to none of B_0, B_1, B_2, in defiance of the fact that $B_0 \cup B_1 \cup B_2 = EG$. In case (3), we have

$$B_i \cap EH \in \{\emptyset, \{a, b, c\}, \{c, f, g\}, \{b, c, d, g\}\}$$

for each $i \in J_2^0$. But for each i we must have $B_i \cap EH \neq \emptyset$, for otherwise $B_0 \cup B_1 \cup B_2 \neq EG$. Therefore, we obtain the contradiction that $c \in B_0 \cap B_1 \cap B_2$. We conclude that no subgraph of G is isomorphic to K_4. A further consequence, from Theorem 9.1, is that $|X| \leq 3$ for any cutset X of G.

We shall show that $|X| = 3$ for any cutset X. Clearly, the choice of G implies that G is 2-connected, for if $G - \{v\}$ had distinct components H_1 and H_2 for

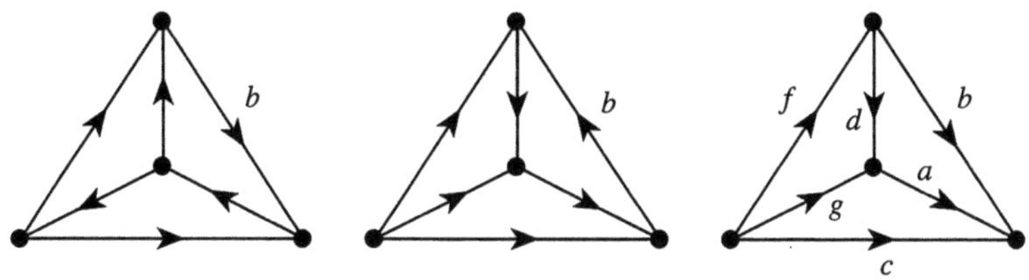

FIGURE 9.1.

some $v \in VG$, then no bond in G could contain edges in both $G[VH_1 \cup \{v\}]$ and $G[VH_2 \cup \{v\}]$. Then the requirement that $EG = B_0 \cup B_1 \cup B_2$ could not be met. Hence $|X| > 1$.

Suppose therefore that $|X| = 2$. Let $X = \{u, v\}$. Since G is chordal, u and v must be joined by an edge h. Since each edge of G belongs to at least one of B_0, B_1, B_2 but not to all three, we may assume without loss of generality that $h \in B_1 - B_0$. For each $i \in J_2^0$, let $B_i = \partial Y_i$, where $u \notin Y_i$. Thus $v \in Y_1 - Y_0$.

Since $G - X$ is not connected, there exist complementary non-empty subsets A and B of $VG - X$ such that $[A, B] = \emptyset$. As $Y_0 \cap X = \emptyset$ (since $u \notin Y_0$ and $h \notin B_0$), we may assume that $Y_0 \subseteq A$. Let $G_A = G - B$ and $G_B = G - A$. For each $i \in J_2^0$ define $Y_i' = Y_i \cap VG_A$ and $B_i' = [Y_i', VG - (B \cup Y_i')]$. Thus B_i' is a bond of G_A. We shall show that $\{B_0', B_1', B_2'\}$ is a 3-ring in G_A.

Each edge of B_i' joins a vertex of $Y_i' \subseteq Y_i$ to a vertex of $VG - (B \cup Y_i') = VG_A - Y_i' = VG_A - (Y_i \cap VG_A) \subseteq VG - Y_i$, and so $B_i' \subseteq B_i$ for each $i \in J_2^0$. Hence $\{B_0', B_1', B_2'\}$ is consistently orientable, and $B_0' \cap B_1' \cap B_2' = \emptyset$. Since $Y_0 \subseteq A$, we have $B_0 \subseteq EG_A$. Moreover, for each $i \in J_2^0$, any edge of $B_i \cap EG_A$ joins a vertex of $Y_i \cap VG_A = Y_i'$ to a vertex of $VG - (B \cup Y_i) = VG - (B \cup Y_i')$ and therefore belongs to B_i'. It follows that $B_0 \cap B_1 \subseteq B_0' \cap B_1'$ since $B_0 \cap B_1 \subseteq EG_A$. Hence $B_0' \cap B_1' \neq \emptyset$. Similarly, $B_0' \cap B_2' \neq \emptyset$. If $h \in B_1 - B_2$, then the same argument with B_0 replaced by B_2 shows that $B_1' \cap B_2' \neq \emptyset$. In the remaining case, $h \in B_1 \cap B_2$, and since $h \in EG_A$, it again follows that $B_1' \cap B_2' \neq \emptyset$. Therefore $\{B_0', B_1', B_2'\}$ is a 3-ring in G_A.

Now suppose that $\{D_0, D_1, \ldots, D_{k-1}\}$ is a k-ring of G_A with $k > 3$ and that $\bigcup_{j=0}^{k-1} D_j \subseteq B_0' \cup B_1' \cup B_2'$. For each j, define $D_j' = D_j$ if $h \notin D_j$, and let $D_j' = D_j \cup (B_1 \cap EG_B)$ otherwise. Then D_j' is a bond in G for each j, and it follows immediately that $\{D_0', D_1', \ldots, D_{k-1}'\}$ is a k-ring of G, in contradiction to the maximality of $\{B_0, B_1, B_2\}$. Hence $\{B_0', B_1', B_2'\}$ is a maximal 3-ring in G_A. This result contradicts the choice of G. We conclude that $|X| = 3$.

Suppose that $|VG| \geq 4$. Since no subgraph of G is isomorphic to K_4, there must be two distinct vertices that are not adjacent. If w is one of them, then some subset of the neighbours of w forms a cutset $\{x, y, z\}$. Hence, we have the contradiction that $\{w, x, y, z\}$ is the vertex set of a subgraph of G isomorphic to K_4.

Thus $|VG| \leq 3$. The lemma follows immediately, as the three bonds in K_3 are not consistently orientable. ■

In preparation for the general case of an n-ring, where $n > 3$, we assert the following self-evident lemma. (See Figure 9.2.)

Lemma 9.6 *Let S and T be subsets of the vertex set of a connected graph G and let $\bar{S} = VG - S$ and $\bar{T} = VG - T$. Then*

(1) $\partial(S \cap T) \subseteq \partial S \cup \partial T$;

(2) *every bond that is a subset of $\partial S \cup \partial T$ and distinct from ∂S and ∂T must meet $\partial S - \partial T$ and $\partial T - \partial S$;*

(3) $\partial S \cap \partial T = [S \cap T, \bar{S} \cap \bar{T}] \cup [S \cap \bar{T}, \bar{S} \cap T]$.

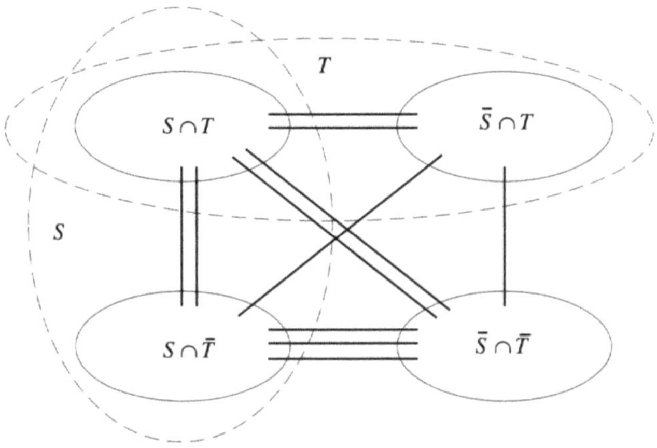

FIGURE 9.2.

Let $\mathcal{B}$ be a strict maximal n-ring $\{B_0, B_1, \ldots, B_{n-1}\}$ of bonds of a connected graph G, where $n \geq 4$ and $B_i \cap B_j \neq \emptyset$ if and only if $i = j$ or $i \equiv j \pm 1 \pmod{n}$. We assume as given an orientation σ of G under which every bond in $\mathcal{B}$ is directed. For each $i \in J^0_{n-1}$, we choose $e_i \in B_i \cap B_{i+1}$. Let $B_i = \partial A_i$ for each i. Since $n \geq 4$, we have $e_{i-2} \notin B_i$. Therefore, we may define D_i to be the element of $\{A_i, VG - A_i\}$ that contains no vertex incident on e_{i-2}. We also write $\bar{D}_i = VG - D_i$. For all i, we define $\xi(i) = 0$ if G is oriented so that every edge of B_i is directed to its end in D_i, and we let $\xi(i) = 1$ otherwise.

The next lemma and its corollary are immediate consequences of Lemma 9.6.

Lemma 9.7 *We have $\xi(i) = \xi(i+1)$ if and only if*

$$B_i \cap B_{i+1} = [D_i \cap D_{i+1}, \bar{D}_i \cap \bar{D}_{i+1}].$$

On the other hand, $\xi(i) \neq \xi(i+1)$ if and only if

$$B_i \cap B_{i+1} = [D_i \cap \bar{D}_{i+1}, \bar{D}_i \cap D_{i+1}].$$

Corollary 9.8 *If $\xi(i) = \xi(i+1)$, then $[D_i \cap \bar{D}_{i+1}, \bar{D}_i \cap D_{i+1}] = \emptyset$. If $\xi(i) \neq \xi(i+1)$, then $[D_i \cap D_{i+1}, \bar{D}_i \cap \bar{D}_{i+1}] = \emptyset$.*

Lemma 9.9 *If an edge of B_j joins two vertices of D_i, then $j \equiv i \pm 1 \pmod{n}$.*

Proof. Suppose that e is an edge of B_j that joins two vertices of D_i, where $j \not\equiv i \pm 1 \pmod{n}$. We shall obtain the contradiction that $\mathcal{B}$ is not strict.

Suppose further that, for every $k \notin \{i-1, i, i+1\}$, either $B_k \subseteq EG[D_i]$ or $B_k \subseteq EG[\bar{D}_i]$. Recall that $e_{i-2} \in EG[\bar{D}_i]$. Since $e_{i-2} \in B_{i-2}$, it follows that every edge of B_{i-2} joins two vertices of $\bar{D}_i$. Since $B_{i-2} \cap B_{i-3} \neq \emptyset$, the same is also true of B_{i-3} (if $n > 4$) and by induction of B_j, in contradiction to the existence of e. Note also that $B_i \cap B_k = \emptyset$, so that no edge of B_k joins a vertex of D_i to a vertex of $\bar{D}_i$. We conclude that, for some $k \notin \{i-1, i, i+1\}$, there exist an edge $a \in B_k \cap EG[D_i]$ and an edge $b \in B_k \cap EG[\bar{D}_i]$.

Note that $\partial(D_i \cap D_k) \subseteq B_i \cup B_k$ by Lemma 9.6(1). Hence, some subset of $\partial(D_i \cap D_k)$ is a bond B such that $B \subseteq B_i \cup B_k$. Since B_i and B_k are disjoint members of $\mathcal{B}$, the strictness of $\mathcal{B}$ will be contradicted if we can prove that $B \notin \{B_i, B_k\}$. But $B \neq B_k$, since b joins two vertices of $\bar{D}_i$ and each edge of B has an end in D_i. It remains to show that $B \neq B_i$.

Suppose that $B_i \cap EG[\bar{D}_k] = \emptyset$. It follows that $\partial(D_i \cap \bar{D}_k) \subseteq B_k$. In fact, this inclusion is strict, since $b \in B_k \cap EG[\bar{D}_i]$. On the other hand, $D_i \cap \bar{D}_k \neq \emptyset$ since $B_k \cap EG[D_i] \neq \emptyset$, and so $\partial(D_i \cap \bar{D}_k) \neq \emptyset$ by the connectedness of G. These results contradict the fact that B_k is a bond. Hence, $B_i \cap EG[\bar{D}_k] \neq \emptyset$. We infer that $B_i \neq B$, as required. ■

The following corollary is immediate.

Corollary 9.10 *For all $j \notin \{i-1, i, i+1\}$, we have $B_j \subseteq EG[\bar{D}_i]$.*

Lemma 9.11 *For each i, $\xi(i) \neq \xi(i+1)$.*

Proof. Assume that $\xi(i) = \xi(i+1)$ for some i, and without loss of generality take $\xi(i) = 0$. We shall contradict the maximality of $\mathcal{B}$ by constructing a ring $\mathcal{B}'$ in which B_i and B_{i+1} are replaced by three bonds: B'_i, B', and B'_{i+1}.

Note that $e_{i-1} \in B_i = [D_i, \bar{D}_i]$ and, by Corollary 9.10, $e_{i-1} \in B_{i-1} \subseteq EG[\bar{D}_{i+1}]$. Thus $e_{i-1} \in [D_i \cap \bar{D}_{i+1}, \bar{D}_i \cap \bar{D}_{i+1}]$. Let D'_i be the vertex set of the component of $G[D_i \cap \bar{D}_{i+1}]$ containing an end of e_{i-1}, and let $B'_i = \partial D'_i$.

We show that B'_i is a bond. Let $G' = G - D'_i$. If L is a component of G' containing vertices of D_{i+1}, then L contains all vertices of D_{i+1} since $G[D_{i+1}]$ is connected and $D'_i \subseteq \bar{D}_{i+1}$. Now $[D_i \cap D_{i+1}, \bar{D}_i] \neq \emptyset$ by Lemma 9.7, since $B_i \cap B_{i+1} = \emptyset$ otherwise. Thus $G[\bar{D}_i]$ is a subgraph of L by Corollary 1.5 and the facts that $G[\bar{D}_i]$ is connected and $D'_i \subseteq D_i$. Therefore, any component M of G' other than L contains only vertices of $D_i \cap \bar{D}_{i+1}$. Since G is connected, we have $[VM, D'_i] \neq \emptyset$, so that the existence of M contradicts the definition of D'_i. Thus L is the only component of G' and it follows that B'_i is a bond.

Similarly, for any $X \subseteq VG$ such that $G[X]$ is a component of $G[\bar{D}_i \cap D_{i+1}]$, ∂X is a bond. Let D be the union of all such X for which $\partial X \cap B_{i+2} \neq \emptyset$.

By Lemma 9.7, $B_i \cap B_{i+1} = [D_i \cap D_{i+1}, \bar{D}_i \cap \bar{D}_{i+1}]$. Hence e_i has neither end in $D_i \cap \bar{D}_{i+1}$, and so $B'_i \subseteq (B_i \cup B_{i+1}) - \{e_i\}$. Thus B'_i is distinct from B_i and B_{i+1} and therefore contains an edge $e'_i \in B_{i+1} - B_i$ by Lemma 9.6(2). In particular, $e'_i \in [D_i \cap \bar{D}_{i+1}, D_i \cap D_{i+1}]$ since $B'_i = \partial D'_i$. Hence, just one end of e'_i is in $D_{i+1} \cap \bar{D}$, where $\bar{D} = VG - D$. Let D' be the vertex set of the component of $G[D_{i+1} \cap \bar{D}]$ containing an end of e'_i, and define $B' = \partial D'$.

We now show that B' is a bond. Since $D' \subseteq D_{i+1}$ and B_{i+1} is a bond, $\bar{D}_{i+1}$ is included in a single component Y of $G - D'$. Choose a component Z of $G - D'$. It suffices to show that $Z = Y$. If $VZ \cap D \neq \emptyset$, then for some vertex set X such that $G[X]$ is a component of $G[D]$, we have $X \subseteq VZ$ since $D' \subseteq \bar{D}$. Because $G[X]$ is a component of $G[\bar{D}_i \cap D_{i+1}]$, any edge $e \in \partial X$ must have an end $w \in D_i \cup \bar{D}_{i+1}$. If $w \in \bar{D}_{i+1}$, then neither end of e is in D'. In this case, $e \in G - D'$, so that $e \in EZ$, and it follows that $Z = Y$ by Corollary 1.5. Hence, we assume that $\partial X = [X, D_i] \subseteq B_i$. But now we have the contradiction that $\partial X \cap B_{i+2} = \emptyset$. On the other hand, suppose that $VZ \cap D = \emptyset$. If $Z \neq Y$, then $VZ \subseteq D_{i+1} \cap \bar{D}$. Thus $G[D']$ could not be a component of $G[D_{i+1} \cap \bar{D}]$ because $[D', VZ] \neq \emptyset$ by the fact that G is connected. This contradiction shows that $Z = Y$ so that B' is a bond.

Suppose that $B' \cap B_{i+2} \neq \emptyset$. Choose $a \in B' \cap B_{i+2}$. Then $a \in EG[\bar{D}_i]$ by Corollary 9.10. By the definition of B', one end of a must be in $\bar{D}_i \cap D_{i+1}$ but not in D. Since $a \in B_{i+2}$, the definition of D therefore implies that $a \in EG[D_{i+1}]$. Thus, both ends of a are in $G[D_{i+1} \cap \bar{D}]$, in contradiction to the fact that $a \in B'$. Hence $B' \cap B_{i+2} = \emptyset$.

In particular, $e_{i+1} \notin B'$. Thus $B' \subseteq (B_i \cup B_{i+1}) - \{e_{i+1}\}$ and $B' \neq B_{i+1}$. Therefore B' contains an edge $e' \in B_i - B_{i+1}$. Then $e' \in [D_{i+1} \cap \bar{D}, D]$. Now we define D'_{i+1} to be the vertex set of the component of $G[D]$ containing an end of e'. Let $B'_{i+1} = \partial D'_{i+1}$. We have already shown that B'_{i+1} is a bond because $G[D'_{i+1}]$ is a component of $G[D]$.

Define

$$\mathcal{B}' = (\mathcal{B} - \{B_i, B_{i+1}\}) \cup \{B'_i, B', B'_{i+1}\}.$$

We now aim to show that $\mathcal{B}'$ is a ring of bonds and thus contradicts the maximality of $\mathcal{B}$.

Let us show that $\mathcal{B}'$ is consistently orientable. Let σ' be the orientation of G such that, for each $d \in EG$, we have $\sigma'(d) = \sigma(d)$ if and only if

$$d \notin B' \cup ((B_i \cup EG[D_i]) \cap (B_{i+1} \cup EG[D_{i+1}])).$$

Recall that $B' \subseteq \partial(D_{i+1} \cap \bar{D})$. By Corollary 9.10, it follows that B' is disjoint from the bonds in $\mathcal{B} - \{B_i, B_{i+1}, B_{i+2}\}$, since each of these bonds is included in $EG[\bar{D}_{i+1}]$. These bonds are also disjoint from $EG[D_{i+1}] \cup B_{i+1}$, and are therefore directed bonds under σ' since they are directed bonds under σ.

We have already shown that $B' \cap B_{i+2} = \emptyset$. Moreover $B_{i+2} \cap EG[D_i] = \emptyset$ and $B_{i+2} \cap B_i = \emptyset$, so that B_{i+2} is a directed bond under σ' as well as under σ.

It follows from Corollary 9.8 that

$$B'_i \subseteq (B_i \cap EG[\bar{D}_{i+1}]) \cup (B_{i+1} \cap EG[D_i]).$$

Recall that each edge of B_i is directed under σ to its end in D_i, since $\xi(i) = 0$. Thus, each edge of $B_i \cap EG[\bar{D}_{i+1}]$ in B'_i is directed to its end in D'_i under σ. It cannot be in $B' \subseteq \partial(D_{i+1} \cap \bar{D})$ or in $B_{i+1} \cup EG[D_{i+1}]$ and is therefore also directed to its end in D'_i under σ'. Similarly, each edge of $B_{i+1} \cap EG[D_i]$ in B'_i is directed to its end in D_{i+1} under σ and therefore to its end in D'_i under σ'. Hence B'_i is a directed bond under σ'.

Next, consider B'. Each edge $b \in B'$ has an end $x \in D'$ and an end $y \in VG - D'$. Thus $x \in D_{i+1}$. If $b \in B_{i+1}$, then b is directed to x under σ, since $\xi(i+1) = 0$. Otherwise $b \in B' - B_{i+1}$. Then $y \in D_{i+1}$, and so $y \in D$ because $b \in B' \subseteq \partial(D_{i+1} \cap \bar{D})$. Hence $y \in \bar{D}_i \cap D_{i+1}$. If $x \in \bar{D}_i$, then $x \in \bar{D}_i \cap D_{i+1}$, and we have the contradiction that x is in D, being in the same component of $G[\bar{D}_i \cap D_{i+1}]$ as y. We conclude that $x \in D' \cap D_i$, and again b is directed to x under σ. Therefore B' is directed under both σ and σ'.

Now choose $c \in B'_{i+1}$. Suppose first that $c \in [D_i, D'_{i+1}]$. Then $c \in B_i \cap (EG[D_{i+1}] \cup B_{i+1})$ since $D'_{i+1} \subseteq \bar{D}_i \cap D_{i+1}$. Moreover, c is directed to its end in D_i under σ and therefore to its end in D'_{i+1} under σ'. On the other hand, if $c \notin [D_i, D'_{i+1}]$, then $c \in [\bar{D}_i \cap \bar{D}_{i+1}, D'_{i+1}]$ since $B'_{i+1} \subseteq \partial(\bar{D}_i \cap D_{i+1})$. Because $\xi(i+1) = 0$, c is directed under σ to its end in D'_{i+1}. Note that $c \notin B' \subseteq \partial(D_{i+1} \cap \bar{D})$, for the end of c in D_{i+1} is in $D'_{i+1} \subseteq D$. Nor is c in $B_i \cup EG[D_i]$. Therefore c is also directed under σ' to its end in D'_{i+1}. Hence B'_{i+1} is a directed bond. We conclude that $\mathcal{B}'$ is consistently orientable.

It remains to be shown that the bonds in $\mathcal{B}'$ can be cyclically ordered so that distinct bonds are disjoint if and only if they are not consecutive. By construction $e_{i-1} \in B_{i-1} \cap B'_i$, $e'_i \in B'_i \cap B'$, and $e' \in B' \cap B'_{i+1}$. Moreover, $B'_{i+1} \cap B_{i+2} \neq \emptyset$ by the definition of D. On the other hand, B'_i, B', and B'_{i+1} are subsets of $B_i \cup B_{i+1}$ and are therefore disjoint from $\bigcup_{j=0}^{i-2} B_j \cup \bigcup_{j=i+3}^{n-1} B_j$. We have also shown that $B' \cap B_{i-1} = B' \cap B_{i+2} = \emptyset$. Moreover, $B_{i-1} \cap B'_{i+1} = \emptyset$ since $B_{i-1} \subseteq EG[\bar{D}_{i+1}]$, $B'_i \cap B'_{i+1} = \emptyset$ by Corollary 9.8, and $B'_i \cap B_{i+2} = \emptyset$ because $B_{i+2} \subseteq EG[\bar{D}_i]$ by Corollary 9.10.

Hence $\mathcal{B}'$ is a ring of bonds. This conclusion contradicts the maximality of $\mathcal{B}$. Therefore $\xi(i) \neq \xi(i+1)$ for each i. ■

The desired result is now immediate.

Corollary 9.12 *Every strict maximal ring of bonds of length greater than* 3 *is even.*

Combining this result with Lemma 9.5, we obtain the following theorem due to Little and Holton (1982).

Theorem 9.13 *Every strict maximal ring of bonds in a graph is even.*

It follows from Lemmas 9.3 and 9.4 and Kuratowski's theorem that every non-planar graph has a strict maximal odd ring of circuits. Conversely, Theorems 9.2 and 9.13 imply that no planar graph has such a ring. Thus, we have another characterisation of planar graphs.

Theorem 9.14 *A graph is planar if and only if it has no strict maximal odd ring of circuits.*

10
Bridges

In this final chapter, we present yet another characterisation of planar graphs, due to Tutte, and also prove a theorem of Whitney on planar imbeddings of simple 3-connected planar graphs.

10.1 Residues and bridges

Let C be a circuit in a planar 2-connected graph G, and suppose that $C \subset EG$. Then some vertex of VC is of degree greater than 2. Accordingly, by Lemma 1.38 and the planarity of G, there is a set $\mathcal{C}$ of circuits in G that constitutes a spanning circuit double cover. Some subset $\mathcal{S}$ of $\mathcal{C}$ satisfies the condition that $\sum \mathcal{S} = C$. Let $\mathcal{T} = \mathcal{C} - \mathcal{S}$. Then $\sum \mathcal{T} = \sum \mathcal{C} + \sum \mathcal{S} = \sum \mathcal{S} = C$. On the other hand, choose a circuit $D \in \mathcal{C}$. Since $\sum \mathcal{C} = \emptyset$, we have $D = \sum(\mathcal{C} - \{D\})$, and so $\mathcal{C} - \{D\}$ also spans $\mathcal{C}(G)$. By Corollary 1.36 it follows that $\mathcal{C} - \{D\}$ is a basis of $\mathcal{C}(G)$. Therefore, there is a unique subset of $\mathcal{C} - \{D\}$ whose sum is C. We may assume that this subset is $\mathcal{S}$, since $\mathcal{S}$ and $\mathcal{T}$ cannot both contain D. If $\mathcal{U}$ is a third subset of $\mathcal{C}$ with sum C, it follows that $D \in \mathcal{U}$. But $\sum(\mathcal{C} - \mathcal{U}) = C$ and $D \notin \mathcal{C} - \mathcal{U}$, so that $\mathcal{C} - \mathcal{U} = \mathcal{S} = \mathcal{C} - \mathcal{T}$. Hence $\mathcal{U} = \mathcal{T}$. We conclude that $\mathcal{S}$ and $\mathcal{T}$ are the only subsets of $\mathcal{C}$ whose sum is C. We call $\mathcal{S}$ and $\mathcal{T}$ the *residues* of C (with respect to $\mathcal{C}$).

An edge $e \in EG - C$ is said to be a *chord* of C if its ends are vertices of VC. In this case, we describe $\{e\}$ as a *bridge* of C. If K is a component of $G - VC$, then $EK \cup [VK, VC]$ is also reckoned as a *bridge* of C. If B is a bridge of C, then the vertices of $VG[B] \cap VC$ are called *vertices of attachment* of B (to C). The other vertices of $VG[B]$ are called *interior* vertices of B.

Choose $e \in B$, where B is a bridge of C. If e belongs to a circuit in $\mathcal{S}$, then e belongs to a second circuit in $\mathcal{S}$, since $e \notin C = \sum \mathcal{S}$. As $\mathcal{C}$ is a double cover, e then belongs to no circuit of $\mathcal{T}$. Similarly, if e belongs to a circuit of $\mathcal{T}$, then e belongs to two circuits of $\mathcal{T}$ and to no circuit of $\mathcal{S}$. In the former case, we call e an *$\mathcal{S}$-edge* and in the latter case a *$\mathcal{T}$-edge*. Thus e is an $\mathcal{S}$-edge if $e \in \bigcup \mathcal{S}$ and a $\mathcal{T}$-edge if $e \in \bigcup \mathcal{T}$. No other cases are possible.

Lemma 10.1 *Let $\mathcal{C}$ be a spanning circuit double cover of a planar 2-connected graph G. Let C be a circuit of G with a bridge B and a residue $\mathcal{S}$. If B has an $\mathcal{S}$-edge, then every edge of B is an $\mathcal{S}$-edge.*

Proof. Suppose not. Then some vertex $v \in VG[B] - VC$ is incident on an $\mathcal{S}$-edge a and a $\mathcal{T}$-edge b, by Corollary 1.9. Since G is 2-connected, we have $\{a, b\} \subseteq D$ for some circuit D. It follows that $|D \cap \bigcup \mathcal{S} \cap \partial v| = 1$, yet the intersection of each member of $\mathcal{C}$ with $\bigcup \mathcal{S} \cap \partial v$ is of even cardinality (2 for members of $\mathcal{S}$ and 0 for members of $\mathcal{T}$). This contradiction to the fact that $\mathcal{C}$ is spanning completes the proof. ■

In the light of Lemma 10.1, we now describe B as an *$\mathcal{S}$-bridge* if $B \cap \bigcup \mathcal{S} \neq \emptyset$ and a *$\mathcal{T}$-bridge* otherwise.

Lemma 10.2 *Let G be a 2-connected planar graph with spanning circuit double cover $\mathcal{C}$. Let C be a circuit in G with residue $\mathcal{S}$. Let P be a path in G, disjoint from C and included in $\bigcup \mathcal{S}$, joining two vertices u and v of VC but having no internal vertex in VC. Let Q be a path included in C and joining u and v.*

(1) *There is a subset $\mathcal{R}$ of $\mathcal{S}$ such that $\sum \mathcal{R} = P \cup Q$.*

(2) *Let $w \in VQ - \{u, v\}$. Then any edge of $\bigcup \mathcal{S}$ incident on w must belong to $\bigcup \mathcal{R}$.*

Proof. (1) Let $\mathcal{T} = \mathcal{C} - \mathcal{S}$. Since $P \cup Q$ is a circuit, it has a residue $\mathcal{W}$. Thus $\sum \mathcal{W} = P \cup Q$. If $\mathcal{W} \subseteq \mathcal{S}$, we are done, and so we suppose that $\mathcal{W} \cap \mathcal{T} \neq \emptyset$. In this circumstance, it suffices to show that $\mathcal{T} \subseteq \mathcal{W}$, for then $\mathcal{C} - \mathcal{W} \subseteq \mathcal{S}$ and $\sum(\mathcal{C} - \mathcal{W}) = P \cup Q$.

If $C \in \mathcal{T}$, then $\mathcal{T} = \{C\}$ since $\sum\{C\} = C$. In this case, $\mathcal{T} \subseteq \mathcal{W}$. We suppose therefore that $C \notin \mathcal{T}$. Then there exists a $\mathcal{T}$-bridge A of C that meets $\bigcup \mathcal{W}$. A does not meet P since P is included in an $\mathcal{S}$-bridge. Therefore A is a subset of a $\mathcal{W}$-bridge of $P \cup Q$ by Lemma 10.1. $\mathcal{W}$ thus contains every circuit of $\mathcal{T}$ that meets A. It remains only to show that every $\mathcal{T}$-bridge of C meets $\bigcup \mathcal{W}$.

There exists a vertex $x \in VC \cap VG[A]$ since G is connected. Let $\mathcal{T}_x$ be the set of $\mathcal{T}$-bridges of C meeting ∂x. Thus $A \in \mathcal{T}_x$.

We claim that every $\mathcal{T}$-bridge in $\mathcal{T}_x$ meets $\bigcup \mathcal{W}$. Let $\mathcal{U}$ be the set of $\mathcal{T}$-bridges in $\mathcal{T}_x$ that meet $\bigcup \mathcal{W}$. $\mathcal{W}$ therefore contains every circuit of $\mathcal{T}$ that meets a bridge in $\mathcal{U}$. Since G is 2-connected and $C \subseteq \bigcup \mathcal{T}$, by Theorem 1.26, G has a circuit D that meets both $\partial x \cap \bigcup \mathcal{U}$ and $\partial x \cap (\bigcup \mathcal{T} - \bigcup \mathcal{U})$. Thus $|D \cap \partial x \cap \bigcup \mathcal{U}| = 1$.

Since $\mathcal{C}$ spans $\mathcal{C}(G)$, it follows that some circuit M in $\mathcal{T}$ meets both $\partial x \cap \bigcup \mathcal{U}$ and $\partial x \cap (\bigcup \mathcal{T} - \bigcup \mathcal{U})$. The former condition implies that $M \in \mathcal{W}$ and therefore that $M \cap \partial x \cap \bigcup(\mathcal{T}_x - \mathcal{U}) = \emptyset$. Hence M meets $C \cap \partial x$. But each member of $\partial x \cap \bigcup \mathcal{U}$ carries exactly two members of $\mathcal{T}$, and so a second circuit in $\mathcal{T}$ must have an intersection with $\partial x \cap \bigcup \mathcal{U}$ of odd cardinality. In other words, this circuit must meet both $\partial x \cap \bigcup \mathcal{U}$ and $\partial x \cap (\bigcup \mathcal{T} - \bigcup \mathcal{U})$ and hence $C \cap \partial x$. Since $\sum \mathcal{T} = C$ and $|C \cap \partial x| = 2$, no other member of $\mathcal{T}$ can meet $C \cap \partial x$. Hence, no circuit of $\mathcal{T}$ can contain exactly one edge of $\partial x \cap \bigcup(\mathcal{T}_x - \mathcal{U})$. But if $\mathcal{T}_x - \mathcal{U} \neq \emptyset$, then there are circuits in G that do so, since G is 2-connected, and therefore we obtain a contradiction to the fact that $\mathcal{C}$ is spanning.

Now choose a vertex $y \in VC - \{x\}$ incident on a member of a $\mathcal{T}$-bridge of C. Let $\mathcal{T}_y$ be the set of $\mathcal{T}$-bridges of C meeting ∂y. It suffices to show that some $\mathcal{T}$-bridge in $\mathcal{T}_y$ meets $\bigcup \mathcal{W}$. Moreover, we may assume that there is a path X included in C that joins x to y and enjoys the property that none of its internal vertices is incident on a member of a $\mathcal{T}$-bridge of C. Since G is 2-connected, some circuit of G meets both $C \cap \partial y$ and $\partial y \cap \bigcup \mathcal{T}_y$. Since $\mathcal{C}$ spans $\mathcal{C}(G)$, at least one circuit, and hence at least two circuits, of $\mathcal{T}$ contain just one edge of $\partial y \cap \bigcup \mathcal{T}_y$ and thus meet $C \cap \partial y$. Since $\sum \mathcal{T} = C$, we infer from the choice of y that some circuit of $\mathcal{T}$ includes X and meets a $\mathcal{T}$-bridge in $\mathcal{T}_x$ and another in $\mathcal{T}_y$. This circuit is therefore contained in $\mathcal{W}$, and we deduce that some $\mathcal{T}$-bridge in $\mathcal{T}_y$ meets $\bigcup \mathcal{W}$, as required.

(2) Suppose not. Let $Z = \partial w \cap (\bigcup \mathcal{S} - \bigcup \mathcal{R})$. Since $\mathcal{R} \subseteq \mathcal{S}$ and $\partial w \cap \sum \mathcal{R} = \partial w \cap \sum \mathcal{S} = \partial w \cap C$, we have $Z \cap C = \emptyset$. It follows that every component of $\mathcal{C}$ has an intersection with Z of even cardinality, yet G, being 2-connected, has circuits without this property since $Z \neq \emptyset$. This contradiction to the fact that $\mathcal{C}$ is spanning proves that $Z = \emptyset$, as required. ■

Lemma 10.3 *Let G be a 2-connected planar graph with spanning circuit double cover $\mathcal{C}$. Let C be a circuit in G with residues $\mathcal{S}$ and $\mathcal{T}$. Let P be a path in G, disjoint from C, joining two vertices u and v of VC but having no internal vertex in VC. Let Q be a path in G, disjoint from C, joining two vertices w and x of $VC - \{u, v\}$ but having no internal vertex in $VC \cup VP$. Suppose that $w \in VC_x[u, v]$. (See Figure 10.1.) If $P \subseteq \bigcup \mathcal{S}$, then $Q \subseteq \bigcup \mathcal{T}$.*

Proof. Suppose that $P \cup Q \subseteq \bigcup \mathcal{S}$. By Lemma 10.2(1), the circuit $D = P \cup C_x[u, v]$ has a residue $\mathcal{R} \subseteq \mathcal{S}$.

By Lemma 10.2(2), any edge of $\bigcup \mathcal{S}$ incident on w must belong to $\bigcup \mathcal{R}$. We infer that $Q \cup C_w[u, v]$ is included in an $\mathcal{R}$-bridge B of D. Let e be the edge of $C_w[u, v]$ incident on u. Since $e \in B$, e belongs to two circuits in $\mathcal{R}$. However, $e \in C = \sum \mathcal{S}$, so that e belongs to only one circuit in $\mathcal{S}$. These results contradict the fact that $\mathcal{R} \subseteq \mathcal{S}$. Thus, if P is included in an $\mathcal{S}$-bridge, then Q must be included in a $\mathcal{T}$-bridge. ■

We can now prove the following well-known theorem of Whitney (1933).

Theorem 10.4 *A simple 3-connected planar graph has a unique spanning circuit double cover.*

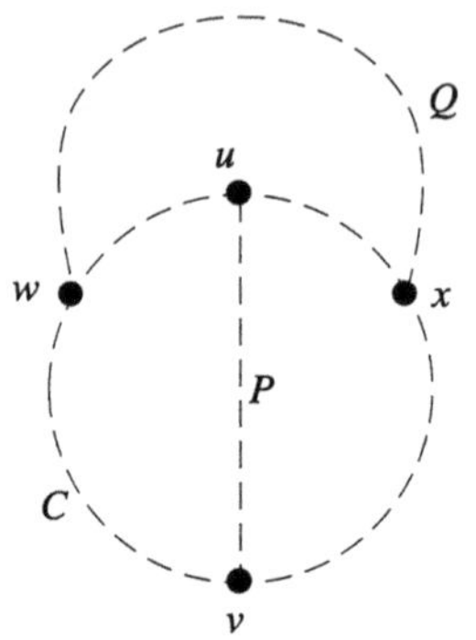

FIGURE 10.1.

Proof. Let $\mathcal{C}$ be a spanning circuit double cover of a simple 3-connected planar graph G. We shall show that the members of $\mathcal{C}$ must be the circuits of G that have a unique bridge.

Let C be a circuit of G with residues $\mathcal{S}$ and $\mathcal{T}$, and suppose that C has a unique bridge. We may assume this bridge to be an $\mathcal{S}$-bridge. Since no $\mathcal{T}$-bridge exists, we must have $\bigcup \mathcal{T} = C$. Hence $\mathcal{T} = \{C\}$, so that $C \in \mathcal{C}$.

Conversely, choose $C \in \mathcal{C}$. Thus C has a residue $\mathcal{S}$ such that $\bigcup \mathcal{S} = EG$. Hence every bridge is an $\mathcal{S}$-bridge. The proof of the theorem will be complete if we can show that C has a unique bridge. Since $G[C]$ is not 3-connected, C certainly has a bridge B. The next step is to show that $VC \subseteq VG[B]$.

Suppose there exists $x \in VC - VG[B]$. We may choose vertices y and z of $VC \cap VG[B]$ that are joined by a path R, included in C, which passes through x and has no internal vertex in $VG[B]$. Then y and z are not adjacent in $G[C]$, by the fact that G is simple if $|B| = 1$ and by the assumption of 3-connectedness otherwise. The latter assumption implies that $G - \{y, z\}$ is connected, and so there must be a path P in G joining an internal vertex of R to a vertex of $VC_x[y, z] - \{y, z\}$ and having no internal vertex in VC. There is also a path Q, included in B, which joins y to z and has no internal vertex in VC (since $G[B] - VC$ is connected). Since P and Q are included in $\mathcal{S}$-bridges, Lemma 10.3 shows that $VP \cap VQ \neq \emptyset$. Hence $P \subseteq B$, a result that violates the choice of y and z. We conclude that $VC \subseteq VG[B]$ for any bridge B. Since G is simple, it follows that $|VG[B] \cap VC| \geq 3$.

Let u, v, w be distinct vertices of VC. There is a path $K \subseteq B$ joining u to v and passing through no other vertex of VC, since $\{u, v\} \subseteq VG[B]$ and $G[B] - VC$ is connected. Since $|VG[B] \cap VC| \geq 3$, we have $|K| > 1$. Hence, there must be a path $L \subseteq B$ joining w to an internal vertex of K and passing through no vertex of $VC - \{w\}$. If we choose L to be of minimal length subject to the above conditions, then we have $|VK \cap VL| = 1$; let b be the unique vertex of $VK \cap VL$. Now suppose that there exists another bridge A of C. By the same argument, we can find a path M, included in A, that joins u to v and passes through no other vertex of VC. M has an internal vertex a joined to w by a path N that is included in A and passes through no vertex of $(VC \cup VM) - \{a, w\}$. (See Figure 10.2.)

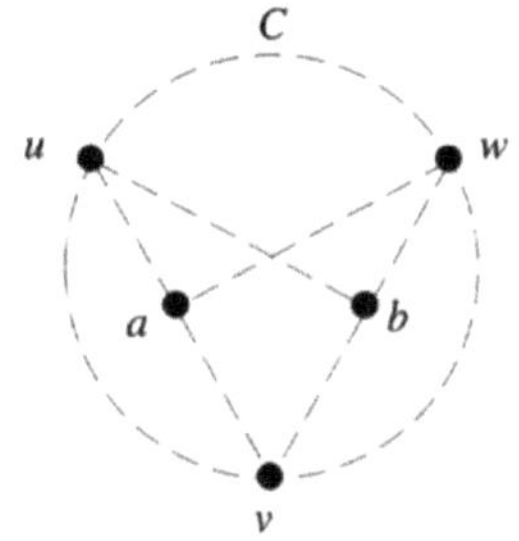

FIGURE 10.2.

By Lemma 10.2, there is a subset $\mathcal{U}$ of $\mathcal{S}$ such that $\sum \mathcal{U} = K \cup C_v[u, w] \cup C_u[w, v]$, and a subset $\mathcal{W}$ of $\mathcal{U}$ such that $\sum \mathcal{W} = K \cup M$. Note that $C_w[u, v]$ is included in a bridge of $D = K \cup M$. Recalling that $\mathcal{W} \subseteq \mathcal{S}$, we see that this bridge cannot be a $\mathcal{W}$-bridge, since each edge of C belongs to only one member of $\mathcal{S}$. Similarly, there is a bridge of D that is not a $\mathcal{W}$-bridge and includes $C_v[u, w] \cup C_u[w, v] \cup L \cup N$. We now have a contradiction, for Lemma 10.3 shows that one of the paths $C_w[u, v]$ and $L \cup N$ must be included in a $\mathcal{W}$-bridge of D. We conclude that C has only one bridge. ∎

The next (rather technical) lemma is akin to Lemma 10.3, which is used in the proof.

Lemma 10.5 *Let G be a 2-connected planar graph with spanning circuit double cover $\mathcal{C}$. Let C be a circuit in G with residues $\mathcal{S}$ and $\mathcal{T}$. Let P be a path in G, disjoint from C, joining two vertices u and v of VC but having no internal vertex in VC. Let P' be a path in G joining an internal vertex x of P to a vertex w of $VC - \{u, v\}$ and having no internal vertex in $VC \cup VP$. Let Q be a path in G, disjoint from C, joining v and w but having no internal vertex in $VC \cup VP \cup VP'$. Let Q' be a path in G joining u to an internal vertex y of Q but having no internal vertex in $VC \cup VP \cup VP' \cup VQ$. (See Figure 10.3.) If $P \subseteq \bigcup \mathcal{S}$, then $Q \subseteq \bigcup \mathcal{T}$.*

Proof. Suppose that $P \cup Q \subseteq \mathcal{S}$. By Lemma 10.2(1), the circuit $D = P \cup C_u[v, w] \cup C_v[w, u]$ has a residue $\mathcal{R} \subseteq \mathcal{S}$. By Lemma 10.2(2), any edge of $\bigcup \mathcal{S}$ incident on w must belong to $\bigcup \mathcal{R}$. Hence P' and $Q \cup Q'$ are both included in

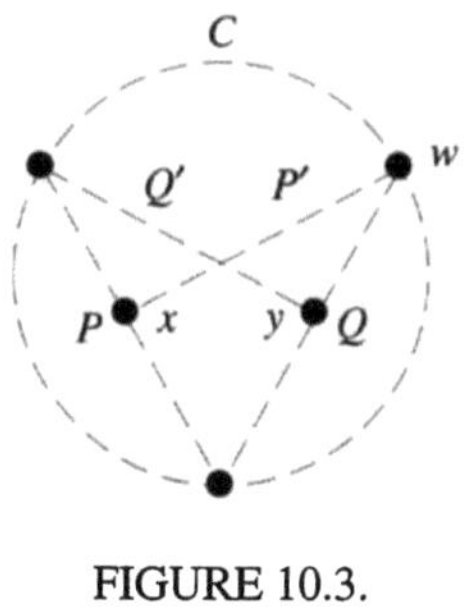

FIGURE 10.3.

$\mathcal{R}$-bridges of D. Thus $P' \subseteq \bigcup \mathcal{R}$ and $Q[v, y] \cup Q' \subseteq \bigcup \mathcal{R}$, in defiance of Lemma 10.3. ■

10.2 Tutte's characterisation

We move on to a characterisation of planarity, due to Tutte (1958), which is expressed in terms of bridges of a circuit.

Let A and B be distinct bridges of a circuit C in a graph G. We say that A *avoids* B if there is a path P, included in C, which passes through every vertex of attachment of A but none of its internal vertices is a vertex of $G[B]$. If this is the case, then none of the internal vertices of the path $C - P$ is a vertex of $G[A]$, and $C - P$ passes through every vertex of attachment of B. Hence B avoids A, and we may say that A and B *avoid* each other. Two bridges of C that do not avoid each other are said to *overlap*. We may construct a simple graph whose vertices are the bridges of C, two vertices being adjacent if and only if they overlap. This graph is called the *overlap* graph of C (in G). Note that if H is the block of G that includes C, then the overlap graph of C in G is isomorphic to a graph obtained from the overlap graph of C in H by the possible adjunction of some isolated vertices.

Again, let A and B be distinct bridges of C. We say that A is *skew* to B if there exist distinct vertices u and v of $VC \cap VG[A]$ such that the paths included in C that join u and v both have an internal vertex in $VG[B]$. In this case, it is clear that B is also skew to A, and so we may say simply that A and B are *skew*.

Note that skew bridges overlap: Any path included in C that passes through the vertices u and v of attachment of A necessarily has a vertex of $VG[B]$ as an internal vertex. Conversely, we have the following lemma. We first define two bridges to be *equivalent* if they have the same vertices of attachment.

Lemma 10.6 *Let A and B be distinct overlapping bridges of a circuit C in a graph G. Then A and B are either skew or they are equivalent bridges with just three vertices of attachment.*

Proof. Suppose that the lemma is false.

We show first that A and B cannot be equivalent. If they were, then $n \neq 3$, where n is the number of vertices of attachment. But if $n \leq 2$, then A and B would avoid each other. Hence $n \geq 4$, and we obtain the contradiction that A and B are skew.

Since A and B are not equivalent, we may suppose without loss of generality that there exists a vertex $x \in VC \cap (VG[B] - VG[A])$. Let P be the path of minimal length that is included in C, passes through x, and joins two vertices y and z of $VG[A]$. Since A and B overlap and the path $C_x[y, z]$ passes through every vertex of attachment of A, some internal vertex of $C_x[y, z]$ belongs to $VG[B]$. Thus, we have the contradiction that A and B are skew. ■

Combining Lemma 10.6 with Lemma 10.3 and Lemma 10.5, we observe that if a 2-connected graph is planar, then the overlap graph of any circuit must be bipartite.

We now prove the converse of this assertion, thereby obtaining the characterisation of planar graphs due to Tutte.

Theorem 10.7 *A graph is planar if and only if the overlap graph of every circuit is bipartite.*

Proof. If G is a planar graph, then each block of G is planar by Lemma 7.2. In view of the preceding discussion, it therefore remains only to show that if a graph G is non-planar, then there is a circuit in G whose overlap graph is not bipartite. In fact, we shall prove by induction on $|EG|$ that G has a circuit C such that its overlap graph includes a triangle or pentagon whose edges join skew pairs of bridges of C. This assertion is true if G is K_5 or $K_{3,3}$: The edge set of the overlap graph of a Hamiltonian circuit is a pentagon in the former case and a triangle in the latter, and in both cases all pairs of overlapping bridges are skew. It follows immediately that the assertion holds if G is a subdivision of K_5 or $K_{3,3}$.

In the remaining case, G has an edge e such that $G - \{e\}$ is non-planar, by Kuratowski's theorem. By the induction hypothesis, $G - \{e\}$ has a circuit C whose overlap graph H includes a triangle or pentagon in which each edge joins a skew pair of bridges. The overlap graph K of C in G includes the same triangle or pentagon unless e joins interior vertices of distinct bridges of C in $G - \{e\}$ and those bridges are vertices of the triangle or pentagon.We may assume in fact that those bridges are adjacent in H, for if they are non-adjacent vertices of a pentagon, then K is isomorphic to the skeleton of the graph obtained from H by identifying them and therefore includes a triangle in which each edge joins a skew pair of bridges.

We infer that there are skew bridges A and B of C in $G - \{e\}$ and that e joins interior vertices of A and B. Thus, there are vertices w and y of attachment of A to C and vertices x and z of attachment of B to C such that w, x, y, z are distinct and one of the paths included in C and joining w to y passes through x while the other passes through z. We may assume w and x to be chosen so that no internal vertex of the path $C_z[w, x]$ is a vertex of attachment of A or B. Since e joins a vertex of $VG[A]$ to a vertex of $VG[B]$, there exists a non-empty path $P \subseteq A$ joining w to an end of e, and there exists a non-empty path $Q \subseteq B$ joining x to the other end of e. Moreover, $C_x[w, z] \cup C_w[z, x] \cup Q \cup \{e\} \cup P$ is a circuit, D. There is a bridge X of D, included in A, which has y and a vertex of $VP - \{w\}$ among its vertices of attachment. A second bridge Y of D is included in B, and has z and a vertex of $VQ - \{x\}$ among its vertices of attachment. A third bridge Z of D includes $C_z[w, x]$ and counts w and x among its vertices of attachment. The choice of w and x implies that any two of X, Y, Z are distinct and hence skew. We conclude that the overlap graph of D in G has a triangle with vertex set $\{X, Y, Z\}$. ■

Bibliography

D. Archdeacon (1980) A Kuratowski theorem for the projective plane, Ph.D. thesis, Ohio State University.

D. Archdeacon (1981) A Kuratowski theorem for the projective plane, *J. Graph Theory* **5**, 243–246.

D. Archdeacon, C. P. Bonnington, and C. H. C. Little (1995) An algebraic characterization of planar graphs, *J. Graph Theory* **19**, no. 2, 237–250.

N. Biggs (1976) On the duality of interaction models, *Math. Proc. Cambridge Philos. Soc.* **80**, 429–435.

A. A. Chernyak (1980) The Little conjecture on planar graphs, *Izv. Akad. Nauk Beloruss. SSR Ser. Fiz.-Mat.* **2**, 41–45, in Russian.

M. Dehn and P. Heegaard (1907) Analysis situs, *Enzyklopädie der mathematischen Wissenschaften III* **AB3**, 153–220.

D. A. Dirac (1961) On rigid circuit graphs, *Abh. Math. Sem. Univ. Hamburg* **25**, 71–76.

J. R. Edmonds (1960) A combinatorial representation for polyhedral surfaces, *Notices Am. Math. Soc.* **7**, 646.

J. R. Edmonds (1965) On the surface duality of linear graphs, *J. Res. Nat. Bur. Standards* **69B**, 121–123.

R. B. Eggleton and D. K. Skilton (1984) Double tracings of graphs, *Ars Combin.* **17A**, 307–323.

M. Ferri and C. Gagliardi (1982) Crystallization moves, *Pacific J. Math.* **100**, 85–103.

J. E. Graver and M. E. Watkins (1977) *Combinatorics with emphasis on the theory of graphs*, Springer-Verlag, New York.

J. L. Gross and T. W. Tucker (1987) *Topological graph theory*, Wiley-Interscience, New York.

A. Jacques (1968) Sur le genre d'une paire de substitutions, *C. R. Acad. Sci. Paris Sér. I Math.* **367**, 625–627.

J. Keir and R. B. Richter, Walks through every edge twice II, *J. Graph Theory*, to appear.

S. Lins (1982) Graph-encoded maps, *J. Combin. Theory Ser. B* **32**, 171–181.

C. H. C. Little (1988) Cubic combinatorial maps, *J. Combin. Theory Ser. B* **44**, 44–63.

C. H. C. Little and D. A. Holton (1982) Rings of bonds in graphs, *J. Combin. Theory Ser. B* **33**, 1–6.

W. McCuaig (1984) A simple proof of Menger's theorem, *J. Graph Theory* **8**, 427–429.

B. Mohar (1987) Generalisation of MacLane's characterization of planar graphs, *Congr. Numer.* **56**, 231–236.

B. Richter and H. Shank (1984) Left-right paths off the plane, *Congr. Numer.* **45**, 115–128.

P. Rosenstiehl and R. C. Read (1978) On the principal edge tripartition of a graph, *Discrete Math.* **3**, 195–226.

S. Stahl (1980) Permutation-partition pairs: A combinatorial generalization of graph embeddings, *Trans. Am. Math. Soc.* **1**, 129–145.

S. Stahl (1983) A combinatorial analog of the Jordan curve theorem, *J. Combin. Theory Ser. B* **35**, 28–38.

S. Stahl (1988) A duality for permutations, *Discrete Math.* **71**, 257–271.

C. Thomassen (1981) Kuratowski's theorem, *J. Graph Theory* **5**, 225–241.

W. T. Tutte (1958) Matroids and graphs, *Trans. Am. Math. Soc.* **88**, 144–174.

W. T. Tutte (1973) What is a map?, *New directions in the theory of graphs*, Academic Press, New York, pp. 309–325.

W. T. Tutte (1984) *Graph Theory*, Addison-Wesley, Menlo Park, Calif.

A. Vince (1983) Combinatorial maps, *J. Combin. Theory Ser. B* **34**, 1–21.

K. Wagner (1937) Über eine eigenschaft der ebenen komplexe, *Math. Ann.* **114**, 570–590.

T. R. S. Walsh and A. Lehman (1972) Counting rooted maps by genus I, *J. Combin. Theory Ser. B* **13**, 192–218.

A. Whitney (1932) Non-separable and planar graphs, *Trans. Am. Math. Soc.* **34**, 339–362.

A. Whitney (1933) A set of topological invariants for graphs, *Am. J. Math.* **55**, 231–235.

List of Symbols

$\cup, \cap, -, +$ union, intersection, complement, and sum, p. 1

$\bigcup X, \bigcap X, \sum X$ union, intersection, and sum of members of X, p. 1

$\emptyset$ empty set, p. 1

$|X|$ cardinality of X, p. 1

$\subseteq, \subset$ subset, and proper subset, p. 1

$\mathcal{P}(Y)$ collection of all subsets of Y, p. 1

$\mathcal{P}_n(Y)$ collection of all $X \subseteq Y$ for which $|X| \leq n$, p. 1

$(x, y), xy, (yx)^{-1}$ ordered pair, p. 1

$X \times Y$ Cartesian product, p. 2

J_n $\{1, 2, \ldots, n\}$, p. 2

J_n^0 $J_n \cup \{0\}$, p. 2

$(x_1, x_2, \ldots, x_n)$ ordered n-tuple, p. 2

$x_1 x_2 \cdots x_n$ ordered n-tuple, p. 2

$(x_n x_{n-1} \cdots x_1)^{-1}$ ordered n-tuple, p. 2

$[x_1 x_2 \cdots x_n]$ set of ordered n-tuples, p. 2

AB concatenation of ordered n-tuples, p. 2

$\mathcal{D}f$ domain of f, p. 2

$f|_X$ restriction of f to X, p. 2

$f \cup g$ union of functions, p. 2

$f[X]$ set of all $f(x)$ such that $x \in X$, p. 2

$p(x, y), q(x, y)$ first component, and second component, p. 2

$\mathcal{P}|_Y$ restriction of a partition, p. 2

(X, f) family, p. 2

$(X, f) \cup (Y, g)$ union of families, p. 2

$(X, f) - (Y, g)$ complement of family, p. 2

$VG, EG, \psi G$ vertex set, edge set, and incidence function, p. 3

$\deg_G(v)$ degree of v in G, p. 3

$[G]$ class of graphs isomorphic to G, p. 4

$G - T$ subgraph $(VG, EG - T, \psi G|_{EG-T})$, p. 4

$G[T]$ subgraph induced by $T \subseteq EG$, p. 4

$G - S$ subgraph $(VG - S, U, \psi G|_U)$, p. 4

$G[VG - S]$ subgraph of G induced by $VG - S$, p. 4

$H \cup K$ union of subgraphs, p. 4

$\partial_G v$ cocycle of v, p. 4

$\partial_G S$ cocycle of set S, p. 4

$c(G)$ number of components, p. 4

$[S, VG - S]_G$ set of edges joining S to T, p. 5

$[\{v\}, \{w\}]_G$ set of edges joining v to w, p. 5

VQ vertex set of circuit or path, p. 5

$P[v, w]$ subpath of P joining v to w, p. 7

C_v subpath of C, p. 9

$T(e)$ fundamental circuit, p. 12

S_H contraction of S in H, p. 13

e_H contraction of edge e in H, p. 13

G_e graph obtained by contracting e, p. 13

Fnd(G) foundation of G, p. 94

$\mathcal{Z}_2$ field of residue classes mod 2, p. 18

$\mathcal{C}(G)$ cycle space, p. 19

$\mathcal{C}^{\perp}(G)$ cocycle space, p. 19

φx image of x under permutation φ, p. 23

$(\varphi x\, \varphi^2 x \cdots \varphi^r x)$ cyclic permutation, p. 23

$\prod_{i=1}^{m} \varphi_i$ product of permutations, p. 23

EM edge set of a map, p. 24

VM vertex set of a map, p. 24

$[M]$ class of maps isomorphic to M, p. 25

M^{ι} imbedding of M, p. 25

$[M, \iota]$ class of isomorphic imbeddings, p. 25

(X, π, φ) oriented imbedding, p. 25

$s(x), s_{\theta}(x), s_{\pi}(x), s_{\theta\pi}(x)$ sides of x, p. 26

$s_{\theta\theta}, s_{\pi\pi}, s_{\theta\pi\theta\pi}, s_{\pi\theta\pi}, s_{\theta\pi\pi}, s_{\theta}, s_{\theta\theta\pi}, s_{\theta\pi\theta}, s_{\pi}, s_{\pi\theta}, s_{\theta\pi}$ sides, p. 26

$\Pi(x, k)$ $(\varphi^{k\iota(x)}\pi x, k\iota(x))$, p. 26

$\Phi(x, k)$ $(\pi x, -k\iota(x))$, p. 27

P^{-} orbit of M^{ι}, p. 28

VI vertex set of imbedding, p. 28

EI edge set of imbedding, p. 28

FI set of faces of imbedding, p. 28

FM set of faces of map, p. 28

v_x vertex containing x, p. 28

e_x edge containing x, p. 28

S_x $\{(x, -1), (x, 1), (\pi x, -1), (\pi x, 1)\}$, p. 28

∂f boundary of face f, p. 29

∂S boundary of set of faces, p. 29

$G(K)$ graph underlying gem K, p. 31

$[I]$ equivalence class of imbedding, p. 34

id identity permutation, p. 35

Ψ_K group of permutations generated by K, p. 35

$K(M)$ group of permutations gnerated by K, p. 35

M^* dual of premap, p. 36

$\mathcal{B}(K)$ set of red-yellow bigons, p. 39

$\mathcal{Y}(K)$ set of red-blue bigons, p. 39

$\mathcal{R}(K)$ set of blue-yellow bigons, p. 39

$r(K)$ number of bigons, p. 39

$\chi(K)$ Euler characteristic, p. 39

$K - [b]$ cancellation of 1-dipole, p. 40

$K - [a, b]$ cancellation of 2-dipole, p. 40

ρ_C^* conjugate of orientation ρ_C, p. 53

$\rho_C(e)$ sense of e, p. 53

ρ^* conjugate of orientation ρ, p. 54

$H(\mathcal{C})$ graph formed from collection $\mathcal{C}$ of sets, p. 56

$S(C_1, C_2)$ essence of $\{C_1, C_2\}$, p. 58

$S(\mathcal{D})$ essence of $\mathcal{D}$, p. 59

$\mathcal{F}(K)$ boundary space, p. 63

$N(C)$ necklace of semicycle C, p. 64

C^* conjugate of b-cycle, p. 65

$I(C)$ side of C, p. 65

$O(C)$ side of C, p. 65

[illegible] nucleus of sextic [illegible], p. 60

$p(\Lambda)$ set of [illegible] edges of [illegible], p. 62

[illegible] p. 64

[illegible] p. 73

[illegible] p. 74

[illegible] p. 74

[illegible]

[illegible] p. 77

[illegible]

[illegible]

[illegible] p. 84

[illegible]

[illegible]

[illegible]

[illegible] points, p. 111

[illegible]

[illegible]

[illegible]

[illegible]

[illegible]

[illegible]

[illegible]

[illegible]

[illegible] angle of [illegible] p. 127

[illegible] p. 141

[illegible] p. 145

Index

MIX
Papier aus verantwortungsvollen Quellen
Paper from responsible sources
FSC® C105338

If you have any concerns about our products,
you can contact us on
ProductSafety@springernature.com

In case Publisher is established outside the EU,
the EU authorized representative is:
Springer Nature Customer Service Center GmbH
Europaplatz 3, 69115 Heidelberg, Germany

Printed by Libri Plureos GmbH
in Hamburg, Germany